SUCKERS

Rose Shapiro has written for newspapers, magazines and medical journals including the *Independent*, the *Observer*, *Time Out*, *Good Housekeeping* and the *Health Service Journal*. She lives in Bristol.

ROSE SHAPIRO

Suckers

How Alternative Medicine Makes Fools of Us All

VINTAGE BOOKS
London

Published by Vintage 2009

2 4 6 8 10 9 7 5 3 1

First published in Great Britain in 2008 by Harvill Secker

Vintage
Random House, 20 Vauxhall Bridge Road,
London, SW1V 2SA

www.vintage-books.co.uk

Addresses for companies within The Random House Group Limited
can be found at: www.randomhouse.co.uk/offices.htm

The Random House Group Limited Reg. No. 954009

A CIP catalogue record for this book
is available from the British Library

ISBN 9780099522867

The Random House Group Limited supports The Forest
Stewardship Council (FSC), the leading international forest
certification organisation. All our titles that are printed on
Greenpeace approved FSC certified paper carry the FSC logo.
Our paper procurement policy can be found at:
www.rbooks.co.uk/environment

Printed in the UK by CPI Bookmarque, Croydon, CR0 4TD

Contents

For Sam

A lesson in folly is worth two in wisdom

Tom Stoppard, *Arcadia*

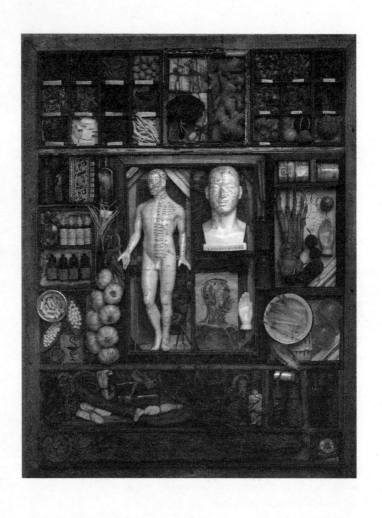

Preface

We are witnessing an epidemic of alternative medicine. There are as many as one thousand different alternative therapies, most with little in common bar one rather important thing: there's no evidence that they work. From chiropractic to colour therapy, reflexology to reiki, such therapies are now used by one in three of us.

Alternative medicine is big business. Latest estimates put the total UK annual spend for Complementary and Alternative Medicine (commonly known as CAM) at £4.5 billion, a market which has grown by nearly 50 per cent during the last ten years.[1] [2] £191m is spent on remedies bought over the counter, such as homeopathic and herbal medicines, but we are increasingly likely to choose a personal consultation with one of nearly fifty thousand alternative practitioners.[3] These self-appointed and predominantly unregulated therapists actually outnumber GPs in the UK. Now even the GPs are getting in on the CAM act – despite their scientific medical training more than half of British GPs now offer alternative medicine, either provided by themselves or by referral.[4]

As with all successful marketing, the world of CAM shows what can be achieved with nothing but a change of name. The same set of practices that was called quackery or fringe medicine in the mid twentieth century was renamed 'alternative medicine' in the 1960s and 70s. The term 'complementary medicine' was coined during the 1990s and now, inspired by the idea that 'alternative' medicine 'can work alongside' and therefore 'complement' orthodox scientific medicine, all these therapies are bundled together as Complementary and Alternative Medicine, or CAM.

Further rebranding has given rise to the notion of 'integrated medicine'. Here's where things are getting out of hand. In a free society every individual has the right to choose to do things they believe are health-promoting as long as they don't directly harm anyone else. But the aim of the integrationists, with Prince Charles as their standard bearer in Britain, is that in future we will all be treated with a mixture of proven, unproven, or disproven therapies and diagnostic techniques, provided by the state. Even more alarmingly, and infuriatingly for fact-favouring sceptics, survey after survey has shown that this approach is what a large proportion of patients want too.

What unproven and disproven therapies do I mean? To make things clear, let's have an alphabetical assortment: acupuncture, Alexander technique, applied kinesiology, aromatherapy, astrological medicine, autogenic training, AuraSoma, Ayurvedic medicine, Bach flower remedies, biochemic tissue salts, biofeedback, bioresonance therapy, biorhythms, Buteyko breathing, chelation therapy, chiropractic, colonic irrigation, colour therapy, cranial osteopathy, craniosacral therapy, crystal healing, ear acupuncture, energy medicine, herbal medicine, homeopathy, Hopi ear candles, holographic repatterning, Indian head massage, iridology, johrei, kombucha tea, light therapy, magnet therapy, metabolic typing, metamorphic technique, moxibustion, naturopathy, nutritional medicine, osteopathy, polarity therapy, radionics, reflexology, reiki, rolfing, shiatsu, therapeutic touch, thought field therapy, urine therapy, vibrational healing and zero balancing. Every one of these either uses diagnostic methods that have no proven, factual basis or involves unsubstantiated or disproven claims of effect and benefit. These remedies and therapies are set to play an increasing role in our health services but oddly, in the era of evidence-based medicine, there is little or no public challenge to their onward march.

Most of these quack treatments are simply ineffectual; such as homeopathy, for example, when there is no medicine in the medicine. Others are dangerous, either in themselves or because they may have been adulterated with uncontrolled amounts of powerful drugs like steroids or banned amphetamines.[5] Many users of alternative medicine defer, or even avoid, having orthodox treatment when they are ill – including people with cancer. A BBC documentary reported the tragic case of a man dying in agony of liver cancer, whose 'diagnostic medical intuitive' healer instructed him (by telephone – she never met him) to shun all orthodox medication, even pain relief.[6]

Nowadays, even when we are not suffering from any specific illness, we are exhorted to strive for something more than the simple absence of disease: a state known as Optimum Health or Absolute Wellness. Optimum Health is a giant undertaking, requiring close reading of special books and newspaper articles, a variety of exclusion diets, internal cleansing practices and the use of a range of therapies from every culture, continent and historical period. Shelf after shelf in expensive wholefood shops like Fresh & Wild is stacked high with such remedies, together with so-called 'intelligent nutrition' supplements. If you didn't know better you would think their wealthy clientele was on its last legs. The opposite is true, of course.

This is a subject about which I feel strongly. What I have learned during the research and writing of this book has only increased my sense of outrage and dismay at the widespread acceptance of alternative medicine and the growing status accorded to it. I mean to show you how alternative medicine puts our health at risk, leaches money and resources from the NHS, is largely unregulated and unaccountable, shortens the lives of people with serious illnesses and makes fools of us all. It is at worst a fraud and when we fall for a fraud our intellectual culture is undermined. We need to understand how and why this has happened, as well as what it says about us, the twenty-first century CAM consumers.

Chapter 1

Ancient and Modern – and how the Me generation became the ME generation

The unique selling point of alternative medicine is that it offers diagnostic systems and therapies that haven't changed in thousands of years, with ancient wisdom offered as the source of its authority. But those who promote alternative medicine have a curiously selective approach to ancient wisdom, frequently reviving some of the least plausible aspects of our ancestors' thinking.

An exploration of the history of medicine immediately exposes a crucial weakness in the philosophy of the CAM campaigners. It shows how they miss one of the most important lessons of history: that it was rejection of superstition and the development of scientific method which brought about a medicine of colossal benefit to humankind. As recently as one hundred years ago people in the West could only expect to live on average until their mid-forties, with infectious diseases killing half of all children before they were ten. Since then, there has been a dramatic increase in life expectancy, evidence that would suggest we are actually healthier than ever before. Now we're likely to live into our eighties and nineties and our governments are beset by pensions crises, debates about retirement ages and the challenge presented by a growing proportion of the population being frail and dependent, plagued by the diseases of old age such as dementia.[7] This new longevity has been made possible by a mixture of scientific medicine and improvements in sanitation, hygiene, living conditions and diet.

Our ancestors weren't so fortunate. Once upon a time a medicine based on faith, magic and superstition was the only medicine. Until around 3000BCE, prayers, incantations and sacrifices were all anyone had. Wherever there were evolving human societies, whether in China, Europe or Australasia, there was an evolving vision of what we now call medicine, but treatment amounted to little more than trial and error. With minimal understanding of even basic anatomy, any recovery would be attributed to the last thing that was tried, to the whim of the gods, to graven images or the position of the planets. Little was understood about the mechanisms of disease. No one realised that a tapeworm could be picked up from eating raw meat or that rabies could be contracted from a wolf bite. If you were ill you simply did or didn't recover, and you often didn't. The unluckiest would be abandoned to die alone or face permanent exclusion from society, like leprosy sufferers were until the emergence of modern treatments in the twentieth century.

We know now that as soon as nomadic hunter-gatherers began to stay put and farm in close proximity to each other and their animals there was bound to be an increase in the incidence of disease. Tuberculosis, flu, measles and even the common cold originated because of this increased contact. Fatal conditions like typhoid, cholera and polio were spread to humans by water polluted by themselves and their stock. And the higher the density of human population, the greater the risk. At that time no one had any idea why, or how to protect themselves from these killer diseases.

Herbal medicine evolved in this early stage of human history. Even if people had no knowledge of pharmacology, they were able to recognise that some roots, berries and leaves had potent effects. The fact that they believed they were exorcising demons rather than administering a medicinal drug – like salicylic acid from the willow tree which was developed into what we now call aspirin – would have made no difference to the outcome. And if you did recover it mattered little whether it was the leaves you ate or the scary dance of the medicine man that was responsible. Survival was proof enough of efficacy.

It is in these earliest days of medicine we first encounter the idea that health is a manifestation of 'balance'. This balance is either within the individual, or between the individual and the whole of nature or even the universe. Illness can therefore only be avoided by living a life of spiritual harmony, in accord with nature and the seasons. Today the idea of 'balance' is back with a vengeance in modern alternative medicine. 'Acupuncture heals imbalances in our health', crystal healing is 'a way of rebalancing physical, emotional and spiritual energies', ear candles 'balance the head on a physical and spiritual level' and 'balance happens when we're in harmony with nature's cues or signals' according to an advert for a light therapy box, designed to treat Seasonal Affective Disorder or SAD.[8] It's a neat and obvious formula with which it is hard to argue. Human beings feel compelled to find patterns in their experience and have always used them, real or imagined, as a way of making sense of the world. As a website promoting Ayurvedic medicine says, 'nature is dualistic: day and night, birth and death, the calm that follows a storm. Harmony is achieved at the midpoint. Health is that state when our mental, physical and emotional aspects are in robust harmony. Ill health is an imbalance but there are simple and natural ways to restore balance'.[9]

It's a short hop from these simplistic notions of balance to the more sophisticated and formal idea of the four humours, the first known description of which was written by ancient Greeks in the fifth century BCE. All early medicine, from Chinese, Egyptian and Persian to Indian is based on this model or something very similar. Humouralism represents a way of thinking about the body, health and disease that was to dominate medicine for the next two thousand years. It is an approach which has survived in some Eastern countries and which still influences the thinking behind much of alternative medicine. The four humours were described as four distinct bodily fluids – yellow bile, black bile, phlegm and blood – linked conceptually to the four elements: fire, earth, water and air. They were mirrored by the seasons: summer, autumn, winter and spring, as well as by the organs of the body: the liver, brain and lungs, gall bladder and spleen respectively. The belief was that all diseases were caused by an imbalance of the humours, called dyscrasia. So, for example, malaria was ascribed not to mosquitoes, but to dyscrasia combined with sultry summer weather.

Hippocrates, who lived in Greece during the fourth century BCE, was the first person to apply these ideas to medicine. But he also believed that the universe and everything in it were subject not to supernatural forces but to physical laws. He thought medicine should therefore be a reasoned application of these physical laws based on a devoted and detailed observation of human patients and their symptoms. Hippocrates had principled moral ideas about doctoring, expressed as a series of pledges including 'to help the sick and never with the intention of doing harm or injury' (rendered in Latin as *primum non nocere* or 'first, do no harm') to maintain patient confidentiality and to share knowledge and expertise with colleagues. These principles are the basis of modern medical ethics and practice.

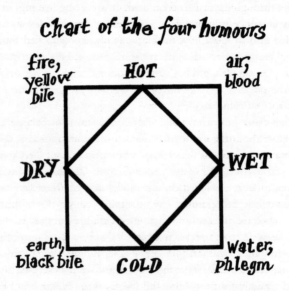

The concept of the four humours would have made sense to ancient peoples because they were aware of only some of the facts. Throughout history our understanding of nature has grown as knowledge expands. We are only able to jettison what turns out to be incorrect in the light of greater knowledge. The world turns out to be round. So it was that gradually, through observation and experiment, we came to understand that mosquitoes carried malaria. Now it's known that only the female mosquito carries the disease and there are currently plans to flood risk areas with millions of genetically modified sterile males in order to reduce the total insect population.[10] How foolish would it be, in the modern world, to cling to the ancient wisdom that suggests that muggy weather plus dyscrasia causes malaria?

Claudius Galen, a Greek born in Turkey in 130CE, embellished humouralism further. After studying medicine for twelve years he became a physician in a gladiator school. This provided him with a crash course in anatomy and he was later to describe wounds as 'windows into the body'. Galen became court physician to Roman emperor Marcus Aurelius when he was able to explore anatomy further through experiments on animals, dissecting live pigs in front of an audience. Cutting the pig's nerves one bundle at a time enabled him to show how the nervous system controlled function: when he cut the laryngeal nerve, now known as Galen's nerve, the pig stopped squealing. They certainly knew how to put on a show in those days.

Galen was able to demonstrate the function of many organs through dissection of both pigs and his favourite subject, the Barbary ape. His experiments convinced him that the mind was located in the brain and not the heart as philosophers argued. But none of the information he gathered through dissection shifted his belief in humouralism. He further refined the idea of the four humours to see them as the defining element of both bodily type and temperament: choleric, melancholic, phlegmatic and sanguine. He showed that the veins and arteries were not full of air, as was commonly believed, but he still thought the venous and arterial systems were separate and he did not realise that blood circulated. He thought blood was created from food and was used up by the body. Excess blood would stagnate in the hands and feet.

Galen still believed that all diseases were caused by dyscrasia, usually brought about by an excess of one of the humours. Bloodletting, emetics and purges were therefore needed in order to expel any harmful surplus. He developed an intricate system to ascertain how much blood should be removed and from where, depending on the disorder, the nature of the patient, the weather and other variables. For centuries blood itself had been imbued with considerable healing power. Historian Roy Porter, in his fascinating medical history *The Greatest Benefit to Mankind* quotes the Roman author

Pliny's report that 'epileptic patients are in the habit of drinking the blood even of gladiators, these persons, forsooth, considering it a most effectual cure for their disease to quaff the warm breathing, blood from man himself, and as they apply their mouth to the wound, to draw forth his very life'.[11]

Galen's twenty-two volumes of medical texts, including the seventeen-volume *On the Usefulness of Parts of the Body*, became the key medical reference work for much of the world. It was translated into Persian and Arabic and many of its concepts formed the basis of Ayurvedic medicine, prevalent in India and increasingly popular today in Europe and the USA. After Galen, the general assumption was that what he didn't know about the body wasn't worth knowing. There were no new studies of anatomy or physiology until the sixteenth century, when Brussels-born Andreas Vesalius realised, through dissecting human cadavers, that human anatomy was different from that of pigs and monkeys.

By then bloodletting had become standard procedure and would be the cornerstone of medicine for hundreds of years. Blood would be let through incisions, scarification or the application of leeches and would stop only when the patient fainted. It is thought that US President George Washington died from a throat infection and shock in 1799 after being drained of nine pints of blood in twenty-four hours. Purging of all kinds was believed to be beneficial, with the widespread use of violent purgatives like calomel (mercury chloride) and tartar emetic.

Such practices were widespread in the West until the end of the nineteenth century when finally it became evident that they had no therapeutic value and were downright dangerous. But bloodletting persists in India, where CNN reports 'every day, about a hundred patients seek out Ghyas' open-air clinic, situated outside Delhi's biggest mosque. To make the blood flow more easily, Ghyas first makes his patients stand in the sun for half an hour. Then, the patient, still standing upright, is tied from the waist down with a rope, and the actual incisions are made with a razor blade. The darker the colour of the blood, the more the patient must bleed.'[12] It's also making a comeback in the West. Some alternative practitioners of traditional Chinese medicine now offer it in the form

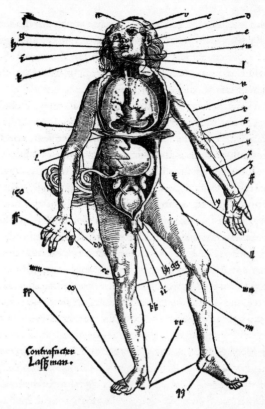

Points for bloodletting from a German medical handbook of 1517

of acupoint bloodletting, achieved by puncturing acupoints with needles. Its various advertised purposes include 'activation of blood, clearance of channels (meridians) and reduction of hotness'.[13]

If such seepage offends, there's always 'dry cupping'. When film actor Gwyneth Paltrow appeared at a New York premiere revealing a back covered in large circular bruises, a BBC report explained how heated cups are applied to the skin 'as a form of acupuncture that focuses on the movement of blood, energy – called Qi – and body fluids, such as lymph, which circulate around the body's tissues. Oriental medicine states that pain is due to stagnation of these systems. Cupping is believed to stimulate flow of blood, lymph and

Qi to the affected area. Its uses include relieving pain in the muscles, especially back pain from stiffness or injury and clearing congestion in the chest, which can occur with colds and flu'.[14]

A belief in humouralism, if not bloodletting, exists unchanged in much of modern 'traditional' medicine, save for an occasional tweak or extra layer that makes it slightly more complicated-sounding. Thus today's Traditional Chinese Medicine is based on a five element model: fire (heart/bitter), earth (spleen/sweet), metal (pungent/lung), water (salty/kidney) and wood (sour/liver). Healthiness requires five element balance. Rather like the playground game of scissors/paper/stone, this has its own circular logic: 'Water controls Fire, while Fire will control Metal. If Water is weak and fails to control Fire, the Fire's action on Metal becomes excessive and Metal is weakened as a result . . . if one element fails to fulfil its controlling/restraining duties then the imbalance can become much more severe and damaging,' as one modern handbook puts it.[15] Ayurveda also proposes five elements – ether, air, fire, water and earth – but only three humours or doshas: wind (vata), choler (pitta) and phlegm (kapha). These are said to act on the seven tissues of the body

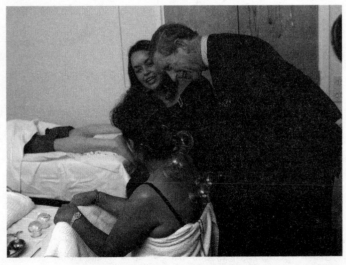

Dry cupping as Prince Charles looks on

(dhatus) and their waste products (malas) as well as three inter-dependent universal constituents, known as The Three Gunas.

European medicine had followed similar lines through the Middle Ages and beyond, incorporating various healing systems, especially the use of herbs, into the humoural approach. Medieval Christian doctrine dictated that God had created a treatment for every ailment, made apparent through God's mark or 'signature' on the remedy. So the spotted leaves of lungwort resemble diseased lungs and were a treatment for tuberculosis; the skull-like seeds of skullcap cured headaches; the maidenhair fern cured baldness and adder's tongue cured snakebite, all early examples of the idea that 'like cures like' which was to manifest itself in homeopathy in the early nineteenth century. The same concept appears in Chinese medicine, where the priapic form of a rhino horn suggested its use as an aphrodisiac. The black rhino is now virtually extinct in the Far East.

The Renaissance brought great advances in knowledge about anatomy. By the turn of the fifteenth century *post mortem* examination of the human body was no longer thought disrespectful to God and the Church began to relax its suppression of medical research and experimentation. Leonardo da Vinci made a huge contribution to the training of doctors with his detailed anatomical drawings, knowledge which helped to improve surgical techniques, such as the tying off of blood vessels after amputation instead of cauterisation with an excruciating mixture of hot oil, water and metal.

By the sixteenth century the Church was the principal provider of health care through infirmaries linked to monasteries. Folk healers, with their herbal remedies and rituals, were excluded. But there was little understanding of infectious disease and the humoural model prevailed, until Swiss-born Paracelsus proposed that agents outside the body could cause illness. Known as the father of toxicology, Paracelsus coined the idea that 'the dose makes the poison', i.e. toxic substances can be benign or even beneficial in small doses; conversely, a benign substance can be deadly if over-consumed.

Infection and the development of scientific medicine

Even though humouralism was beginning to lose its grip, some pretty strange ideas about the nature of disease persisted. Miasma theory was the most widely believed concept of infection. First described by Abaris the Hyperborean in ancient Greece, the idea was that gases from underground decomposition rise up into the air we breathe to cause illness, most notably cholera. Miasma theory was the basis of the new public health movement that took off in the first half of the nineteenth century, supported by campaigners such as Florence Nightingale and Sir John Simon, who became the first Medical Officer of Health for London. And though miasma theory was wrong, it led to improvements in sanitation which went a long way to containing the risk of infection, an interesting example of the right result emerging from the wrong answer.

At the same time the new germ theory emerged. This was the idea that human infection is caused by an external living organism, or *contagium vivum*. In the 1850s two doctors – John Snow working in London and William Budd in Bristol – found strong evidence for germ theory as an explanation of cholera and typhoid epidemics. It wasn't until the 1860s that the French chemist Louis Pasteur was able finally to demonstrate the existence of pathogenic (disease-causing) organisms, followed in the 1880s by the German doctor Robert Koch who finally identified *vibrio cholerae* as the cause of cholera.

Discoveries like this were central to the development of modern scientific medicine. The second half of the nineteenth century saw huge advances, with Germany becoming the world centre for medical research, training and the creation of pharmaceuticals. Through the twentieth century modern scientific medicine duly became increasingly effective. Sophisticated diagnostic techniques and surgical methods have further revolutionised health care and the UK in particular made a huge advance in equality of access when the National Health Service came into being in 1948, with the promise of providing free medical care for all, paid for by taxation.

In the 1970s the scientific approach to medicine was given new impetus by the development of the idea of Evidence Based Medicine, or EBM, 'the conscientious, explicit and judicious use of current best evidence' in making decisions about the care of individual patients.[16] EBM has brought about a sea change in clinical practice, which is now expected to be based on a continuous process of research and testing to show a treatment has demonstrable effect and benefit. In Britain provision is moderated accordingly by the National Institute of Health and Clinical Excellence (NICE).

You might say that it is a strange kind of progress that leads to where we are today, with self-inflicted conditions like obesity and smoking-related diseases becoming the biggest killers in the developed world. Critiques of modern health care, including the role of the profit-hungry pharmaceutical industry, are many and various. Perhaps we expected too much of science and of scientists. Not every treatment works and medicine can kill as well as cure. Supporters of CAM attack the over-use of treatments including surgery, which, they say, has led to unacceptably high levels of iatrogenic (medically-caused) illness and death. Hospital-acquired infections, resistant to antibiotic treatment, kill thousands of vulnerable patients in the UK every year.

But the fact remains that in the developed world we are living longer, healthier lives than we have ever lived before. Seventy is the new fifty and most of us can expect to lead lives largely free of serious illness, which end only after a relatively short period of infirmity in old age. Yes, today more of us die from cancer and heart disease, but a hundred years ago there were a host of other illnesses – notably infectious diseases – which used to carry us off much earlier in our lives.

When advocates of CAM press the case for alternative diagnostics, medicines and therapies they invariably criticise conventional medicine but it is historically absurd to ignore or minimise its achievements and its positive impact on human life. Infection, which used to be the biggest killer, is now countered by vaccines and antibiotics. Few in the West have seen a person die from an infectious disease. Though we now live long enough to contract illnesses like cancer, the survival rates of many types are improving year on year. Destructive new illnesses like HIV/AIDS can in many cases be restrained by drugs. Women have much greater control over their fertility and the risks of childbirth and infancy are negligible compared with those of the past.

This necessarily brief exposition of medical history, primitive and inaccurate notions of anatomy, illness and cure, ought to persuade anyone that human life has been vastly improved by modern scientific medicine. That certainly seemed to be the general view from the end of the nineteenth century onwards as most people turned away from quacks and 'snake oil' merchants. There remained naturopaths like Benedict Lust (1872–1945) who never accepted germ theory, calling it 'the most gigantic hoax of modern times'. Lust believed that illness was caused by 'denatured, devitalised foods, gluttony, sensuality, high heels, corsets, drugs, tobacco, alcohol, over-clothing, over-work, worry, fear, anxiety'. The 'healthy body does not allow undue multiplication of germs,' according to this advocate of the curative powers of year-round sunbathing and the benefits of Porous Health Underwear.[17]

Though there were few takers for such ideas, in the 1950s the American Food and Drugs Administration (FDA) still saw the need to make a public information film presented by actor Raymond Massey, TV's 'Dr Leonard Gillespie', warning people not to buy quack cures, including an oblong object that would supposedly emit 'Z-Rays' to treat arthritis, and taped music that was claimed to cure all types of cancer.[18] The quacks were down, but not out.

The fall and rise of the alternative empire
By the mid twentieth century fringe medicine in the UK was moribund. A few 'health food' shops sold supplements and some herbal remedies. Hardly any alternative practitioners remained, beyond the occasional osteopath serving the minority prepared to pay for treatment outside the free National Health Service. Only faith healers stayed popular and they usually offered their services free through Spiritualist churches.

In *Nature Cures,* James Whorton charts this decline and fall in an account of the prevalence of homeopathy in the USA.[19] At the turn of the twentieth century there were as many as ten thousand homeopaths working in the USA, with seventy-nine homeopathic hospitals and twenty-two homeopathic medical schools. A combination of the 1910 Flexner report, which set higher scientific standards for US medical training, and a drop in enrolment numbers meant that by 1923 only two homeopathic colleges survived.[20] By the 1950s both had closed their doors. Whorton quotes the President of the Connecticut Homeopathic Medical Society saying in 1948 that 'the precipitous drop in the popularity of homeopathy is a frightful phenomenon to behold'. By the early 1970s there were fewer than one hundred practising homeopaths in the United States.

But in the middle of the twentieth century something began to change. The youth culture of the 1960s with its questioning of every authority, including medical, was a driving force. Joni Mitchell in her anti-war, pro-counterculture anthem *Woodstock* sounded a call to connect with a natural world that her generation felt was being destroyed by the greedy and soulless modernity of their parents. The hippies were looking beyond both scientific rationalism and established religion for a different sense of what life was for. These ideas were expressed in 'holism', the theory that matter and reality are made up of organic or unified wholes greater than the simple sum of their parts. This is the root of the now constantly-repeated claim that alternative practitioners look at the 'whole person' and that narrow-minded orthodox doctors address only symptoms and specific body parts, as if they existed in isolation. There was also a growing public awareness of dishonest and corrupt practices within the pharmaceutical industry, which was increasingly perceived as putting profit before patients. Not surprisingly, this contributed further to a generalised distrust of orthodox medicine.

Just as important were the new ideas about preventive medicine and health promotion. The role of what was now called 'lifestyle' in determining health and mortality had become apparent, with the US Surgeon General suggesting in 1979 that half of all deaths were due to unhealthy behaviour. It was up to the individual to seize responsibility by giving up smoking, taking up exercise and eating right. Individualism would take us forward: we were bang in the middle of what the film-maker Adam Curtis calls 'the century of the self' and already well acquainted with the practice of setting personal goals in the quest for happiness and fulfilment.[21] Self-help was the order of the day and a healthy and long life could be attained simply by deciding that we wanted one and by making the necessary 'lifestyle changes'.

Then came something harder to quantify: a pervasive sense of feeling not quite as fit and happy as we expected to feel, despite the triumphs of modern medical science. We were doing better but feeling worse, as the American psychiatric professor Arthur J Barsky put it in his revelatory *Worried Sick*, published in 1988: 'We are disturbed by minor ailments, haunted by the possibility of sickness and plagued by a seeming necessity for constant medical attention . . . we fear not just for our health but for our physical safety in general, since disease is only one of the corporeal threats to which we feel so vulnerable.'[22] Not being ill was no longer enough to make us feel well. We knew deep down we weren't sick, but we were haunted by an uncomfortable sense of what alternative practitioners were starting to call 'dis-ease'.

What we hungered for was a state of Extreme Wellness, also known as Optimum Health. 'Health is wholeness – wholeness in its most profound sense, with nothing left out and everything in just the right order to manifest the mystery of balance' writes Andrew Weil, virtually the CEO of alternative medicine today in America.

'Far from being simply the absence of disease, health is a dynamic and harmonious equilibrium of all the elements and forces making up and surrounding a human being.'[23] The language is all too familiar, as if nothing had been learned in two thousand years: order, mystery, balance, elements and forces.

Ancient medicine, new bottles: CAM now

Today these ideas are ubiquitous and the use of alternative medicine is growing at an amazing rate. So who's most likely to use it? Every survey confirms that in the West it is women in the highest socio-economic group, aged between thirty-five and fifty-five. Middle-aged mothers in the moneyed and educated ABC1 group, the top of the tree in socio-economic classification terms, are most ardent CAM users, with as many as 47 per cent saying they've used it at least once. A World Health Organisation review shows the same pattern of use in every developed country, with middle-aged, middle-class educated and/or wealthy women always constituting the largest user group. Around half the population of European countries like France, Belgium, Germany and Denmark have used CAM and in Australia, America and Canada up to 70 per cent of the population does, so you're unusual if you don't. The WHO review found that the global market for CAM is worth over $60 billion. In America expenditure on CAM remedies is currently $2.7 billion per year. In the UK, Canada and Australia the remedy market is worth $2.3 billion, $2.4 billion and $80 million respectively.[24]

CAM is often used alongside scientific medicine. An average of around 35 per cent of cancer patients use alternative therapies and may not always tell their orthodox doctors, according to a survey of fourteen European countries.[25] An American study suggests that in the US it's as many as 70 per cent of cancer sufferers. This survey found that herbal medicines and remedies were the most commonly used, together with homeopathy, vitamin and mineral supplements, medicinal teas, spiritual therapies and relaxation techniques.[26] As we will see later, alternative remedies can pose a serious problem when they interfere with orthodox treatment. In North America and Europe more than half of all people with HIV/AIDS are also active users of CAM.

While many of those with serious or chronic illnesses such as arthritis, skin problems or breathing difficulties turn to alternative medicine, its keenest consumers see it as the answer to a range of much less clearly defined conditions. Typically, these include allergies, anxiety, back pain, fatigue, depression, digestive problems, food sensitivities, headaches, insomnia, skin disorders, stress and overweight. Conventional medicine rarely fulfils the needs of these patients – described as 'heartsink' by some GPs – who appreciate the time and attention they can buy from a CAM practitioner.

Most strikingly, half of CAM users say they use it when they are not actually ill. Unlike conventional medicine, CAM is something you can engage with regardless of your health status.[27] Even if you have nothing wrong with your back, for example, you can sign up for a 'spinal check' and a course of 'maintenance adjustments' from a chiropractor, or for monthly acupuncture sessions in order 'to prevent future problems by keeping your energy in balance'.[28]

Alternative medicine provides a place in which to express a sense of both 'worried sickness' and of entitlement to treatment and attention. You can be confident that a homeopath will never say it's just a virus and suggest you go home and take a painkiller.

But CAM offers something more potent, in the form of a set of beliefs and practices that can substitute for religious faith. As the writer Mark Lawson suggests, 'just as religious doctrine can make life more tolerable by offering scientifically unverifiable promises that bring psychological benefits, so does alternative medicine'.[29] Extreme Wellness is the nearest we will get to experiencing a state of grace; a detox will deliver us from evil. And as participation in organised religion dwindles, the alternative practitioner has appropriated the caring, listening role of the parish priest.

Not surprisingly, the number of alternative practitioners has increased hugely to meet these demands. In the UK in 2005 they were thought to number around 47,000, a figure made more meaningful when you discover there are only approximately 35,000 GPs in this country. One in ten of us visit a complementary practitioner every year, amounting to twenty-two million consultations.[30] UK accident and emergency departments see around fourteen million people each year. Only osteopaths and chiropractors are legally obliged to have undertaken approved training courses and to be registered with a professional body. Outside these two disciplines, CAM is unregulated. Anyone can decide to set themselves up as a practitioner offering any other type of CAM consultation, diagnostic method or treatment. There are professional bodies for acupuncture, herbal medicine and homeopathy, but membership is voluntary and these organisations have no power to regulate what their members actually do. So whilst prospective customers may be reassured to see initials such as IACBP after a practitioner's name, the International Association for Colour Breathing Practitioners (*sic*) has no legal or regulatory status.

In alternative medicine credentials are often not what they seem. Anyone can obtain a degree, even a PhD, via a non-accredited internet correspondence course. How about a 'Master of Arts in Holistic Wellness' or a 'Doctor of Philosophy in Natural Health'? The virtual colleges offering these sound plausible enough: there's the Australasian College of Health Sciences (confusingly based in Portland, Oregon) or the Clayton College of Natural Health in the US which claims to have more than twenty-five thousand graduates and offers a doctorate at a cost of $5,300 (nearly £3,000), online or by post. This is where Gillian McKeith of *You Are What You Eat* TV and book fame obtained her title 'Doctor'. Of 'holistic nutrition'.[31] [32]

In the UK there has, however, been a belated recognition that patients do need some protection in this environment. A House of Lords report in 2000 recommended that the government should invest in CAM research and that there was a need for more public information about 'what does or does not work and what is or is not safe in CAM'. Five years on came a government move to set up a register of those acupuncturists and herbal medicine practitioners who have done an approved training course. But with no suggestion of an assessment of whether either acupuncture or herbal medicine are clinically effective, or that other types of CAM might need regulation, it was hardly the 'crackdown on bogus alternative medical practitioners' the Department of Health claimed.[33]

The government has also given a £900,000 three-year grant to the Prince of Wales' Foundation for Integrated Health to help initiate 'voluntary self-regulation of other complementary professions'. Once again, without any requirement of efficacy, it's hard to see the value of such measures beyond the application of a veneer of respectability to CAM therapies and their practitioners. But clinical effectiveness is not what is on the mind of the integrationists. Prince Charles gave an idea of their priorities when he launched the Foundation for Integrated Health's five-year strategy in 2003. 'I actually believe the integration of the best of ancient and modern approaches can be of benefit in the relief of unnecessary suffering on the part of patients – that's what I really fundamentally believe and in certain cases can be enormously effective and can save money and that's, I think, a very important feature it seems to me in the whole equation in today's world.'[34] Speaking to GPs a couple of years later Prince Charles put it a little more clearly in his description of a 'whole person' approach to health care which 'makes both orthodox and complementary treatments available to the patient, not as competing or separate approaches, but in unison'.[35]

This, in essence, is what the Foundation for Integrated Health and other CAM organisations are aiming for in the UK health service, particularly within primary care. They argue that it will save money. A 2005 Foundation report led by Christopher Small-wood, a former adviser to Barclays Bank, claims that the NHS would make a £3.5 billion 'economy-wide' saving by offering manipulation therapies such as chiropractic on the NHS. Another £480 million could be slashed from the national drugs bill if 10 per cent of GPs offered homeopathic prescriptions for 'everyday conditions, particularly asthma'.[36] Mr Smallwood may not know this, but patients don't even need to get their prescription filled. Some homeopaths claim they only have to write down the name of the remedy to transmit its benefit to the patient. It's a win–win situation.

There was widespread criticism of the Smallwood report. Professor Edzard Ernst is Britain's first professor of complementary therapies and chair of a CAM research department at Peninsula Medical School Exeter and has undertaken a huge amount of scientific investigation of alternative medicine. He called the Smallwood report 'deeply flawed' and said it was 'based on such poor science it's just hair-raising'.[37] The editor of the medical journal *The Lancet,* Richard Horton, warned that lives would be lost if asthma patients were switched to homeopathic remedies. 'We are losing our grip on a rational scientific medicine that has brought benefits to millions, and which is now being eroded by the complicity of doctors who should know better and a prince who seems to know nothing at all,' he said.[38]

CAM's principal argument for integration is that people really like it and it makes them feel better. Of course feeling better is a good thing, but that does not mean that the use of alternative medicine has any effects on the course of a disease. For as a *British Medical Journal* report points out, patients' positive feelings are not necessarily dependent on an improvement in their condition. 'In one UK survey of cancer patients, changes attributed to complementary medicine included being emotionally stronger, less anxious, and more hopeful about the future even if the cancer remained unchanged.'[39] So we arrive at a point where the fact of whether or not a treatment is clinically effective has become irrelevant. 'Patient satisfaction and demand is providing strong evidence of the efficacy of complementary medicine,' says the British Complementary Medicine Association, 'such that science requires a new paradigm, which recognises the holistic concept'.[40] In other words demand itself constitutes evidence and no other proof need be sought; if science cannot show an alternative method to be safe or effective there must be something wrong with science.

This environment enables homeopaths, for example, to feel free to promote themselves with a catalogue of unsubstantiated claims. Promoting 'Homeopathy Awareness Week' in 2007, the Society of Homeopaths said, with no supporting evidence, 'Homeopathy is great for children! From minor problems such as colic and teething, to allergic disorders such as asthma and eczema, and even complex emotional ailments such as confidence issues and attention deficit disorder, homeopathy is a safe and effective option'.[41]

But there are now stirrings of resistance in the UK. Michael Baum, professor emeritus of surgery at University College London, has for decades been a brave and unswerving opponent of alternative medicine that, as he says, 'places itself above the laws of evidence and practises in a metaphysical domain that harks back to the dark days of Galen'. Professor Baum's open letter to Prince Charles, entitled 'With respect your highness, you've got it wrong', received attention in 2004 despite being dwarfed by the fulsome promotion of CAM in most of the British media, where naturopaths, nutritionists and reiki masters are given free rein to write and say whatever they like.[42] Several pro-science rationalists engage in regular 'quackbusting', notably David Colquhoun, professor of pharmacology at University College London, who has compiled an invaluable internet 'Improbable Science Page'.[43] John Diamond wrote only a few chapters of *Snake Oil*, a compelling critique of complementary medicine, before his death in 2001.[44] Dr Ben Goldacre in his 'Bad Science' column in the *Guardian* and on the internet regularly debunks pseudoscience and the unsubstantiated claims of Gillian McKeith and other recipients of unaccredited correspondence college PhDs.[45] In 2006 and 2007 a group of thirteen scientists wrote an open letter to all NHS trusts urging them to stop funding unproven or disproven alternative therapies. This letter, and the accompanying publicity, are thought to have led some trusts to consider stopping the referral of patients to the five NHS homeopathic hospitals.

In the USA defenders of scientific medicine are faced with similar challenges. With the support of the US National Council Against Health Fraud, retired psychiatrist Stephen Barrett edits the Quackwatch website, whose purpose is to 'combat health-related frauds, myths, fads, and fallacies' with a primary focus on 'quackery-related information that is difficult or impossible to get elsewhere'.[46] Dr Barrett is a staunch critic of the US government-backed National Center for Complementary Medicine (NCCAM) whose real purpose, he believes, is 'to promote pseudoscientific methods ... the project was deliberately crafted to avoid critiquing the methods themselves'.[47] NCCAM has yet to provide evidence in support of any alternative therapy, despite spending \$1 billion on research.

In an unrelated development, magician and leading American sceptic James 'The Amazing' Randi has offered a one-million-dollar prize 'to anyone who can show, under proper observing conditions, evidence of any paranormal, supernatural or occult power or event'.[48] The claims of alternative medicine, whether those of homeopathy, acupuncture or reflexology, all come within the terms of Randi's challenge since they have no rational mechanism of action. No one has yet secured the cash.

Unfortunately these initiatives have had little impact on the public desire for a return to irrational thinking and superstition. But this regressive movement, with its denial of history and of fact, can still be halted. Dr Yasuhiro Suzuki of the World Health Organisation has said that alternative medicine is a victim of both uncritical enthusiasts and uninformed sceptics.[49] It's time to inform the sceptics and to convince the enthusiasts to start doing some critical thinking.

Chapter 2

How to Spot a Quack

If it walks like a duck and it talks like a duck, you can be reasonably sure it is a duck. Promoters of quackery and health fraud can be harder to identify. It is especially confusing when, as is so often the case, they offer a package composed of contradictory diagnostic or therapeutic methods.

The Oxford English Dictionary defines a quack as someone who is an 'imposter in medicine' or 'one who professes a knowledge or skill concerning subjects of which he is ignorant'. The word is derived from the Dutch 'kwakzalver': one who prattles or boasts about their supposedly healing salves.[1] It's more complicated than that though. Such definitions 'suggest that the promotion of quackery involves deliberate deception, but many promoters sincerely believe in what they are doing', Dr Stephen Barrett of Quackwatch has found.[2] But because they are likely to be offering the same type of diagnostic and treatment methods the difference between the two is largely academic – quackery is quackery regardless of the merits or otherwise of the practitioner's character.

The task of spotting quackery is made easier once you know that there are large areas of medicine in which it is never found. These are the ones that involve outcomes which are easily measurable, or where there is a risk of imminent death. There is no homeopathic contraceptive, for example. Nor will reflexology be used following a stabbing, or Chinese herbs in the treatment of acute conditions like a broken leg, appendicitis or heart attack. CAM practitioners know better than to compete with mainstream doctors in the provision of trauma care. As one American CAM provider says: 'If you got hit by a truck there's nothing better than the modern medical system'.[3]

Instead, alternative practitioners are mostly to be found diagnosing and treating chronic conditions such as back pain, fatigue, and food intolerance or when, as one Bristol CAM provider describes it, people 'feel unwell in themselves but are not "ill" in the Western sense'.[4] It's usually quite hard to measure what they do. When they claim to be able to 'rebalance energies' or 'correct yin/yang imbalance at the cellular level' it becomes difficult to measure diagnostic accuracy and whether or how a treatment has worked.[5] [6] Serious evaluation becomes virtually impossible when ideas like 'holism', 'treating the whole person' or 'total health through inner balance' are deployed.[7]

The quack mindset and modus operandi are difficult to pin down partly because the language of CAM, with its lavish use of words like natural, balance, energy, healing and wellness, is abstract and subjective. Take this example from a crystal-healing manual: 'it has long been known that our bodies need to keep in step with their own natural vibrations. Otherwise, we show signs of distress that can all too easily become disease. Crystals, too, vibrate at different frequencies. Thus, if we can find a crystal that shares our own vibrations it will help restore them to their correct frequencies'.[8] These assertions may make sense to a practitioner but they are nonsensical, unproven and ultimately meaningless. 'Vibrations' which exist only in the imagination of the crystal therapist have no demonstrable powers, however alluring the idea.

When they say 'homeopathy is a form of medicine with a 200 year history of the gentle and effective treatment of physical and emotional ailments' it's impossible to know what claims are actually being made. Does 'treatment' mean the same as 'cure'? Even when the language sounds carefully composed, poetic even, the underlying beliefs and exact nature of the proposed treatment may remain a mystery. 'Living beings have an innate ability to achieve a perfect state of balance in body, mind and spirit . . . Reiki may be given for any illness, stress, and injury for an overall feeling of well-being as a complementary therapy without any side effects . . . during Reiki sessions the client remains fully clothed and Reiki can be either hands-off or hands-on the body, the choice is always yours.'[9] In the circumstances I might go for 'hands-off'.

Like teenage slang, the vocabulary of CAM is subject to the vagaries of fashion and often unintelligible. The word 'harmony' was big in the 1980s but is now passé and has been replaced by 'balance'. Words with delusions of scientific grandeur are used to confer credibility on dubious practices. In his invaluable guide to distinguishing science and pseudoscience, physics professor Rory Coker describes how 'pseudoscientists often attempt to imitate the jargon of scientific and technical fields by spouting gibberish that sounds scientific and technical'.[10] So it is that 'Thermo Auricular Therapy' trumps plain old 'ear candling'. The current must-have handbag of a word in alternative medicine circles is 'quantum'. In physics this means the smallest amount of a substance that can exist independently, used mainly in relation to a discrete quantity of energy so small it is indivisible. Quantum theory deals with the behaviour of energy and matter at this level, and has become the 'undisputed champion of mysterious fringe science,' says academic neurologist and sceptic Stephen Novella in his article 'Quacks use quantum mechanics to make themselves look smarter'. 'Gurus like Deepak Chopra, author of *Quantum Healing*, use the word "quantum" to give their philosophy a sciencey feel,' says Novella.[11]

Other CAM practitioners appear to employ 'quantum' without having the faintest idea of its meaning. They use it to suggest that you can in some way determine or change reality – particularly the state of your health – at a cellular level, through the power of your mind, sometimes known as the 'bodymind'. It sounds very important and complicated. 'The body is a quantum mechanical device and quantum healing is healing the bodymind from a quantum level [in] a process of peacemaking wherein one mode of consciousness – the mind – corrects mistakes in another mode of consciousness – the body'.[12] 'Quantum-Touch allows you a dimension of healing that heretofore has not seemed possible. Remarkably, this work amplifies the effectiveness of a wide spectrum of healing modalities.'[13] You may also be interested to know there's recently been 'a breakthrough in automatic dishwashing' in the form of the Quantum Powerball.

But while the belief systems and methodology of alternative medicine practitioners may be many and various, there are a number of common identifying features which, once familiar, make the identification of quackery surprisingly straightforward. Much is owed in what follows to physics professor Robert Park, author of *Voodoo Science* and to Dr Stephen Barrett of Quackwatch, who have identified many of these common elements, which amount to quackery's 'red flags'. Quack spotting is easy once you know the signs.

The disclaimer

We start with what usually, almost as if it is an afterthought, comes at the end of nearly all dubious health claims or pseudoscientific information in newspapers, magazines and the internet. It will go something like this:

WARNING!!! The following information is not intended as a substitute for the advice provided by your own physician or health care provider for any health problem or disease. Before following any of these recommendations you should consult your GP about any medical problems or special health conditions. You should not use the information as a substitute for professional medical advice when deciding on any health-related regimen, including but not limited to diet or exercise. Featured products are not intended to treat, cure or prevent any disease.

The disclaimer is the equivalent of the small print in financial adverts, warning that the value of your investment may go down as well as up and is usually the only truthful statement on the page. It is designed to protect both the author and the publisher from legal action when the CAM method fails to work or if a serious condition is left medically unattended because of bad advice. The disclaimer may also involve a passing critique of the shortcomings of science: 'please note that this bioresonance method has not yet been ratified by today's scientific establishment but 40 years of anecdotal evidence supports its use'.[14] Or it may be constructed in order to cover every angle. Stephen Russell, better known as 'The Barefoot Doctor', has been an *Observer* advice columnist, written handbooks designed for both 'Modern Lovers' and 'Urban Warriors' and has lent his moniker to a line of cosmetics. Russell's BarefootDoctorWorld website, described as an 'online surgery and spiritual realisation school', features a legal disclaimer of nearly five thousand words – that's around fourteen A4 pages.[15]

The universal diagnosis

Holistic therapies are said to be able to treat 'the whole person' rather than the symptom or outward manifestation of disease. And because many alternative medicines claim to treat the whole of you, they can by definition treat everything. Or nothing in particular. 'Holism's' universality can be taken a step further, to the point where some alternative therapists diagnose the same syndrome in every patient they see.

Conversely, they may promote a single remedy or therapy, promised to address a wide range of disparate conditions, making it a truly universal panacea. Confused? Harley Street therapist Paul Lennard puts it this way: 'As a whole body therapy, craniosacral therapy may aid people with almost any condition, by raising the vitality and enabling the body's own self-healing processes to be utilised'. He claims to be able to treat, in alphabetical order, twenty-five different conditions including acne, bronchitis, herpes, learning difficulties, migraine and varicose veins. He ends with W, for 'Whatever you suffer from'.[16]

Chiropractic is perhaps the best and most consistent exemplar of the universal diagnosis in CAM, concentrating as it does on correcting 'subluxations' of the spinal joints. The subluxation is a condition of the spine known only to chiropractors, in which they diagnose individual vertebral joints as being misaligned and moving either too much or too little – you're damned if they do and damned if they don't. Daniel D Palmer, the American grocer turned magnetic healer who invented chiropractic believed that 'a subluxated vertebrae (*sic*) . . . is the cause of 95 per cent of all diseases . . . the other five per cent is caused by displaced joints other than those of the vertebral column.'[17] Although modern chiropractors today mostly concentrate on treating back pain, many still push the subluxation line. 'If spinal bones get stuck and don't move right, they can irritate or chafe delicate nerves . . . this can interfere with the vital "life force" transmitted over your nervous system . . . distorted communications between your brain and your body can cause all kinds of health problems. Subluxations are serious!' according to information distributed in clinics in Britain and America.[18]

The single diagnosis is closely allied to the idea of a single cause for all illness, popularised by American naturopath and émigré 'Dr' Hulda Regehr Clark, author of two popular texts in alternative medicine: *The Cure for All Diseases* and *The Cure for All Cancers*. Hulda Clark has a very different approach, but again it is tightly focused. 'No matter how long and confusing is the list of symptoms a person has, from chronic fatigue to infertility to mental problems, I am sure to find *only two things wrong* [her emphasis]: they have in them pollutants and/or parasites.'[19] Like the UK's *You Are What You Eat* Gillian McKeith, Clark's doctorate was obtained from the Clayton College of Natural Health in the US.

Clark is in no doubt about the cause of cancer. 'All cancers are alike. They are all caused by a parasite. A single parasite! It is the human intestinal fluke. And if you kill this parasite, the cancer stops immediately . . . this parasite typically lives in the intestine where it might do little harm, causing only colitis, Crohn's disease or irritable bowel syndrome, or perhaps nothing at all. But if it invades a different organ, like the uterus, kidneys or liver, it does a great deal of harm. If it establishes itself in the liver, it causes cancer!'[20] Both the universal diagnosis and the universal treatment appear repeatedly in alternative medicine and are unmistakable signs of quackery.

It's based on centuries-old ancient wisdom and the method has not changed over time

It is curious to see what a virtue 'ancient wisdom' has become, especially considering the nature of many ancient cures. Whilst it is true that a substance's relative safety might be suggested by the fact that people have used it for centuries, that doesn't necessarily mean it works. Sometimes the claims are breathtaking. 'In these pages you will learn the health secrets that allowed our ancestors to live long, disease-free lives' writes Jordan Rubin, author of *The Great Physician's Rx for Health and Wellness*.[21]

There is added appeal if a CAM method has its roots in a distant country. The so-called 'ancient healing art' of Traditional Chinese Medicine (TCM) sees the most frequent suggestion that being ancient and unchanging might be desirable qualities in medicine.

It's what you might call the 'long ago and far away' factor. Acupuncture is 'a system of healing which has been practised in China and other Eastern countries for thousands of years . . . the skill of an acupuncturist lies in their ability to make a traditional diagnosis from what is often a complex pattern of disharmony,' says the British Acupuncture Council. 'Chinese tradition dates the use of herbal remedies back to around 3000 BC' writes Penelope Ody in her book on Chinese Herbal Medicine. Today 'these same remedies still form the basis of many Chinese prescriptions that have continued in an almost unbroken tradition for 5,000 years.'[22] You would never know that TCM was fashioned in the twentieth century, as we will see, from a ragbag of therapies in post-revolutionary China.

Some promote a beguiling pic'n'mix of old and new. Never mind the contradictions or how it's done. Deepak Chopra, author of *Quantum Healing*, 'merges the ancient wisdom of Ayurveda with knowledge of modern medicine to create an integrated path to health and wellbeing' at his Chopra Centre, while a Kabbalistic health adviser combines 'ancient wisdom from the east and west with modern cutting edge human and personal development technology'.[23] [24]

Feeling worse is a sign of getting better

Truly the 'Emperor's new clothes' element in alternative medicine, this is how alternative medics can always deal with dissatisfied consumers. Usually described as a 'healing crisis', this is such a widespread idea that it has become what could be termed the sine quack non of alternative medicine. Just as many religions involve self-abasement or suffering in order to reach a higher state of spiritual development, alternative medicine often requires consumers to feel a good deal worse in the quest for Optimum Health.

It's usually put something along these lines: 'Many people who start taking JC Tonic report that after being on it for a few weeks or months they become ill . . . when they get sick they become confused and discouraged . . . if you experience what feels like an illness after using JC Tonic, chances are it is not a sickness, but a healing crisis . . . a process in which the body is overcoming ill health and becoming healthier and stronger'.[25]

This 'bad is good' doublethink is compelling. The purveyors of Flora Balance pills explain what you can expect after consuming the 'soil-based organisms' therein. You'll feel worse, but in a better way. 'Symptoms of the healing crisis may at first be identical to the disease it is meant to heal. But there is an important difference: elimination. A cleansing, purifying process is underway and stored wastes are in a free-flowing state . . . sometimes pain and symptoms during the healing crisis are more intense than that of the chronic disease, but it is temporary and necessary.'[26] Most detox regimes tell users to expect to feel pretty rough during the process with symptoms like spots, nausea, bad breath or headaches. These are said to be a good sign and show the detox is working; like a malign departing spirit, the unspecified toxins are leaving the body.

The idea of a healing crisis is also central to homeopathy, which calls it an 'aggravation'. Considered as a positive sign, some homeopaths say it is likely to occur in about 70 per cent of patients.[27] All bases are covered: the *similar* aggravation is when your presenting symptoms get worse. The *dis-similar* aggravation produces new symptoms and your general health declines. Yet another perplexing variation comes in the form of 'accessory symptoms', when the patient is improving in health and vitality but new symptoms begin to appear.[28]

Physicist Jay Shelton says this effect is an illusion. He points out that in mainstream medicine 'very few other types of interventions cause the primary symptoms to worsen as a helpful and desirable part of the cure; fevers don't increase, tumours don't grow, headaches don't worsen, bacterial counts don't increase because of non-homeopathic treatments directed at the specific complaint'. Shelton argues that statistically, the 'good sign' aspect often *appears* to be true because most patients get better after getting worse. 'If the worsening of symptoms was really a good sign, then patients who initially get worse should ultimately be more likely to get well, get well faster, or stay well longer, or have less intense symptoms than patients who just get well without getting worse first,' he says.[29] Homeopaths have never produced any evidence that homeopathic aggravations exist beyond conjecture based on subjective observation.

Neither were they found when British researchers compared the effects of homeopathy to placebo.[30]

A powerful establishment is said to be suppressing the discovery

There's a cure for your condition, but your doctors won't tell you. They don't want you to have the information because it will undermine their authority or do them out of a job. Not only are they hiding cures, doctors are deliberately making you ill. Or sometimes it is profit-mad capitalism that's to blame, usually in the form of what alternative medicine calls 'Big Pharma', the pharmaceutical industry.

In the UK there is an organisation called What Doctors Don't Tell You. Its information service claims to help people to make informed health choices about what works and what doesn't in both orthodox and alternative medicine. But What Doctors Don't Tell You betrays its bias when it describes its monthly bulletin as a 'review of conventional medicine and safer alternatives'. Alarming tales of deaths from prescription drugs like beta blockers, the risks of surgery and hospital errors are set alongside unquestioning articles promoting CAM, with headlines like 'Treating Crohn's naturally'. In common with so many alternative medicine advisers, What Doctors Don't Tell You sells dietary supplements through its own mail order company. It also provides an alternative medicine practitioner database so people can find their nearest reflexologist, homeopath or reiki master.[31]

Kevin Trudeau, convicted fraudster and self-published American author of the best selling *The Natural Cures 'They' Don't Want You To Know About* has got it all sussed. He advises everyone to stop taking prescription drugs. 'The drug industry does not want people to get healthy. The drug industry wants people to buy more drugs. Healthy people don't need drugs. If everyone in America was healthy, the drug industry would be out of business.' Now readers are queuing up to complain on Amazon feedback sites about the lack of any cures in his book, saying it's just an extended advert for Trudeau's subscription-only website, 'your alternative to drugs and surgery' (lifetime membership $499).[32]

It is no surprise to find that the internet is full of false claims that the cures are out there, even for cancer, but that we are intentionally being left to suffer. 'Since the 1920s, more than 100 *natural* treatments for cancer have been developed that are far superior to surgery, chemotherapy and radiation. Every one of these treatment plans, which yield better cure rates and less pain, have been brutally suppressed' according to Cancertutor.com.[33] This is CAM versus what is characterised as the 'Cancer Industry' – that is, any person or body involved in orthodox treatment of, or research into, cancer. As another campaigning altmed website has it, 'when it comes to corruption and dirty tricks, it's hard to find a better example than the cancer industry, which thrives on the disease much like a tumour. One of the most popular ploys for maintaining their power and control is to scare people away from complementary therapies . . . that are far more effective than chemotherapy, radiation or surgery, none of which have been scientifically proven to increase the lifespan of anyone with cancer.'[34] Tales of spineless, self-serving and lying doctors are legion.

Dr Stephen Barrett confirms there has never been any evidence of a plot to withhold cures from the public and that doctors generally prosper by curing diseases, not by keeping people ill. 'When polio was conquered, iron lungs became virtually obsolete, but nobody resisted this advancement because it would force hospitals to change,' he says. Dr Barrett also describes how CAM misleadingly presents its opposition to medical science as a 'philosophical conflict' or 'paradigm shift' rather than a clash between proven and unproven, disproven or fraudulent methods. 'Paradigm' itself is a favourite word in alternative medicine and has the function of making every way of looking at the world appear as valid as every other. Rejecting 'the biomedical paradigm' is so much more impressive than saying you are simply not interested in evidence.

The evidence for their discovery is anecdotal and supported only by testimonials

Personal testimony is not permitted in scientific articles or other approved evaluations of drugs or medical procedures, as it is a practice open to bias and abuse. It would be all too easy to round up satisfied patients who offer positive reports about a chosen treatment or procedure in letters or interviews.

Quacks may not go in for presenting research evidence for efficacy, but they love to give a simple cure rate, which is frequently given at around 80 per cent – not too high to be thoroughly unbelievable, but high enough for the needy to find irresistible. So success statistics for the Grounding Mattress Pad, 'literally a life-changing technology', are that '85% [of users] reported falling asleep faster; 93% slept better through the night; 100% reported waking feeling more rested in the morning and 74% experienced less chronic back and joint pain'.[35] No details of the user group are given – not even the number of people surveyed. Such assertions could easily be fabricated and the patients don't even have to exist. It is considered evidence enough if a practitioner reports, for example, that 'thousands of people have had life-changing experiences with [colon] cleansing. No wonder they want to tell the world about their success'.[36] Even if you might be persuaded by these strangers' colon experiences there is no way of telling whether there are just as many thousands of dissatisfied or even dead consumers whose testimony goes unrecorded.

Patient testimonials substitute for evidence throughout CAM. They are usually headed 'Success Stories', 'How I got my life back' or 'What our customers say' and tell exhilarating tales of lives enhanced and prolonged. 'I now feel brilliant. Everything has improved. Since last June I have had no relapses. The tightness in my spine that indicates the presence of MS has gone.'[37] 'In the three weeks leading up to my [cancer of the bladder] operation, I consumed 100ml of Colloidal Silver per day. The day before the operation, they performed a camera search inside of my bladder. The search revealed that the tumour had reduced significantly in size. The surgeon said to me that whatever I had been doing, continue doing it, "that is truly a miracle".'[38] Such statements, attributed to an anonymous and mystified medic, represent an interesting subset of the testimonial genre.

Practitioners who have cured themselves offer the ultimate in testimonials. 'Here I am, living proof!' writes Dr Lorraine Day in support of her naturopathic cancer programme, which she says cured her breast cancer.[39] Examples of such claims, made especially in relation to the dramatic disappearance of tumours, abound. Australian Robyn Welch, self-styled 'medical intuitive', is a telephone healer – meaning she claims to heal people via the telephone, not that she mends faulty handsets. Welch appeared on a BBC2 documentary exposé *Psychic Surgeon*: 'Robyn, a lady d'un uncertain age, sports a very particular sort of new age bling – silver lipstick and frosted highlights offset by the sort of pricey-yet-tasteless pastel leisurewear favoured by blowsy pier-end fortune-tellers,' as the *Observer*'s Kathryn Flett neatly put it.[40] [41]

As a child Robyn Welch rescued six injured hens and returned them to health: 'Little did I realise that I was tapping into the soul side of my earthly program and they were my first clients.' She only became aware of her special healing ability when she was diagnosed with a uterine tumour in 1979. 'For four days [Welch] concentrated white light used as a directed ray to the affected area. When admitted to hospital, tests showed it had disappeared. In 1981, when she was recovering from a car accident . . . Robyn used the same method to heal herself again.' Today Welch claims to be able to see, via the telephone, inside clients' bodies in order to 'remove negative interference' and replace 'run down circuitry'. The body 'is then entered to work on areas discovered during the field trip, using the life force as a tool similar to a laser ray which is driven by generated heart energy . . . surgery is performed if necessary'. Welch is thoroughly conversant with latest scientific theory and in publicity for a 2006 'Cutting Edge Healing Seminar' she promised that 'the topic of quantum physics will be explored as this is the zone or dimension I attune [*sic*] to the success of my work.'

As *Psychic Surgeon* reported, Robyn Welch was charging thousands of pounds per course of treatment. MRI and PET scanners are pretty much redundant when Robyn's on the line with 'negative interference in the energy field' appearing to her 'like television snow patches'. She says 'these areas relate to parts of our body and unless that "snow" is cleared, it will explode in the body as an illness'. Welch claims a 90 per cent success rate, which of course is unsubstantiated.[42] She was filmed on the telephone to a six-year-old girl conducting an animated yet distinctly one-sided conversation with the girl's internal organs, addressing them as 'Captain Pituitary', 'Thelma Thyroid' and 'Larry Liver'.[43] Telephone healing enables Welch to have a worldwide client list and for years she had nothing to worry about but time zones. The BBC film resulted in widespread derision in the UK press of Welch and her activities.

Brandon Bays, the American author of *The Journey* that has spawned an international network of personal empowerment and healing groups, is associated with a similar tale of impressive self-help. 'Diagnosed with a tumour the size of a basketball, Brandon was catapulted on a remarkable, soul-searching and ultimately freeing healing journey in which she uncovered a means to get direct access to the soul, to the boundless healing potential inside all of us. Only six-and-one-half weeks later she was pronounced tumour free, text-book perfect, no drugs – no surgery!! "Through the humbling and profoundly transformational experience of naturally healing from a tumour in only six-and-one-half weeks, I uncovered a boundless joy and freedom that have been my daily experience ever since. This is the most priceless gift of my life".'[44] Price-less maybe, but not price-free. In the UK a 'Journey Intensive Weekend' in which you can 'uncover old cell memories, clear them completely, letting healing begin' costs £245. A further week's training in order to become an 'accredited practitioner' costs £2,003, but that does include what she calls a 'Manifest Abundance Retreat'.

Sometimes the credit for a cure is shared with a higher being. 'This is an incredible gift coming from God through you Robyn to me' read one ME sufferer's testimonial to Robyn Welch and 'I thank God for giving me another chance and for the best Christmas present I've ever had'.[45]

Very, very rarely, one reads an account of failure. In March 2000 an American cat, by the name of Kitty, was diagnosed with throat cancer and given six months to live. Kitty was given large doses of Cancell, an alternative cancer treatment designed to treat both humans and animals, but sadly she had to be put down at the beginning of July. Despite the outcome her owner's faith in Cancell was unshaken: 'I believe in your product, but the patient has to be willing to live'.[46]

They are flattering and appeal to your vanity

Quacks are keen to stress your individuality, how you are unique and extraordinary and not like other people. You may be encouraged to believe that even if a remedy is useless for others it might still work for you. Or you might be told that the practitioner has seen and helped many people with a condition just like your own, but that your case is special and more complex requiring a longer and therefore more expensive course of treatment. As one UK osteopath says 'the question most patients ask is "how long is it till I will be well again?" This is the most difficult question for any practitioner to answer. Especially since [chronic fatigue] presents or shows itself in so many different ways, with no patient being exactly the same as another, a little like snowflakes'.[47] This individualising approach fuels the idea that anecdote and testimonial have as much if not more weight than scientific evidence (or the lack of it).

You may be exhorted to 'stand up and take charge of your health,' to 'think for yourself,' to 'be your own doctor' or 'your own health detective, open your own health file and redress any obstacles to the fullest possible health,' again encouraging you to lend greater credence to personal experience than to medical evidence and expert advice.[48] [49] Or you may be invited to develop a siege mentality in order to take up arms (or loofahs perhaps, in the example that follows) against a hostile world: 'Justifying the time to soak in a warm, scented bath without feeling guilty is vital to our long-term mental and physical well-being. We need to establish a strong, anchoring sense of self in order to face the world outside with all its challenges,' says Belinda Grant Viagas in her book *Stress – Restoring Balance to our Lives*.[50] It's one short, dripping step from the scented bath of anchored self-worth to the glistening tiled wall of the classic alternative medicine conversation-stopper: 'it works for me'.

It sounds too good to be true

Here we are really spoilt for choice. How about HeightMax™? Designed 'for 12 to 25 year old height and health conscious consumers only' this is 'our revolutionary, trademark, patent-pending nutrition supplement that is taking the nation by storm. It is the only all-natural, completely safe nutrition supplement that enhances your height potential to the maximum your genetic composition will allow.' The HeightMax™ company has been heavily fined for making bogus claims for the product, even featuring a fictitious specialist and product inventor to appear in their advertising.[51] Or look out for Johanna M Hoeller, a 'no touch' chiropractor based in Seattle who corrects spinal irregularities by simply grunting and clicking her wrists a few inches above the patient – she never makes physical contact.[52]

Why not try, for just £28 in Bristol, a simple reflexology foot massage which 'improves nerve function, calming the mind and releasing emotions; removes tensions from muscles and joints and improves all over circulation; improves efficiency of the digestive system, strengthens the immune system and balances the functions of internal organs, glands and control systems of the body'?[53] And there's

Solving personal problems with Traditional Chinese Medicine

'CancerControl', a vitamin supplement said to cure 97 per cent of cancer patients, most within three days. This product claimed to have FDA approval – not, as it turned out, the Food and Drugs Administration, but the lesser known 'Fighting Diseases Association', invented by CancerControl's promoter[54] who is now serving 17 years in jail for his crimes.

You should also expect that a cure for, say, cancer is available from only one (usually internet-based) source and that all supplies have to be purchased direct from the manufacturer, such as 'Medical Research Products (MRP) of Miami, Florida' which 'is the only manufacturer of Cantron', an alternative cancer treatment.[55]

Also too good to be true, as Professor Robert Park writes in his excellent article 'The seven warning signs of bogus science', is the revolutionary scientific discovery made by a lone genius. While it may be a staple of Hollywood science-fiction films it is hard to find examples in real life, says Park, where 'scientific breakthroughs nowadays are almost always syntheses of the work of many scientists'.[56] Such breakthroughs will appear first not on the internet or at a press conference, but in scientific or medical journals, with the research having been open to peer review before publication.

Evidence of such communal endeavour in CAM is extremely rare. Instead, beware statements such as 'this innatehealth.com website is the product of one researcher's lone work over many years to find a scientific answer to the question of why we are still getting sick'. Warren Ward, the aforementioned lone researcher, based in Wales, is surrounded by a veritable forest of red flags. He claims his discovery amounts to 'a new science' and that 'for the first time in history, the aetiologies [causes or origins of disease] of many chronic conditions are fully explained'. These conditions, which Warren Ward says include asthma, cellulite, diabetes and even MRSA 'can be prevented or treated using simple substances', in this case a product he has named 'ActivSignal', which are small pellets of sodium chloride or common salt. Now why didn't I think of that?

The quackery quotient – ear candling

Not all variations of CAM exhibit every one of these red flags and some will boast more than others. But for the sake of example, let's assess the QQ (quackery quotient) of a practice that has become increasingly popular in the last few years – ear candling. Start counting those red flags.

Lynne Hancher offers ear candling (as well as reflexology, aromatherapy, reiki, Indian head massage and spinal touch) 'in and around Wolverhampton and the West Midlands', UK. She gives us the standard ear candling shtick. Thermal Auricular Therapy, as she prefers to call it, 'is an ancient and natural therapy handed down by many civilisations'. The ancient Greeks apparently kicked the whole thing off, but there are claims that it originated as long ago and far away as Atlantis. 'However the practice reached the modern world via the native American Hopi Indians of North Arizona . . . ancient wall paintings show their importance as initiation rituals and healing ceremonies . . . the candles are still made today on the basis of the old traditional formula.'[57]

Ear candles are made of linen or cotton soaked in paraffin or beeswax. An assortment of natural substances may be added to the wax, such as honey, sage, St John's Wort, periwinkle, jojoba, camomile, rosemary or beta-carotene. The linen is rolled into a narrow, hollow cone of around 22cms in length, looking like a slender and extended brandy snap. If you're lucky there will be a filter at one end, designed to catch the hot wax.

The patient lies down on one side and Lynne puts on some relaxing music. 'The candle is then gently placed into the auditory canal (the ear), where it's lit. You will experience a pleasant crackling sound as the candle burns down.' This is (probably) not your hair catching fire but the sound of the flame, which enables the hollow candle 'to work on a chimney principle, drawing any impurities to the surface where they can be gently removed' attests Lynne. Some people find ear candling so pleasant and relaxing that they fall asleep during the procedure. Afterwards, clients often describe a 'relieved feeling' when the candle is blown out and removed. The therapist will usually break it open in a bowl of water and announce that the visible brown residue it contains is principally earwax that has been sucked out, together with toxins or signs of infection. 'Most people cannot believe what has been collected – *expect to be amazed*!' as another ear candler warns.[58]

What is ear candling for? 'Though results may vary from patient

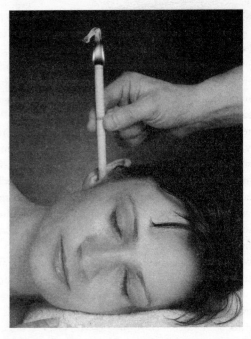

'You will hear the pleasant hissing and crackling of the flame' –
ear candling

to patient' the conditions the candles are supposed to address include 'excessive or compacted wax in the ears, irritation in the ears or sinuses, sinusitis, glue ear, colds, flu, headaches, migraine, stimulation of local and reflex energy flow, energetic revitalisation in cases of hearing impairment, stress, tinnitus', says Lynne. She goes on to say they are 'suitable for most conditions'. Other Thermal Auricular Therapists claim the candles can clean out yeast build-up, promote lymphatic circulation, improve the immune system, stabilise emotions, discourage candida, fortify the central nervous system, purify the blood and open (or indeed align) the chakras.

Do they work? Elsewhere on the internet testimonials come almost as thick and fast as the earwax. 'It makes cleaning the kids' ears fun and they love seeing the YUCKS that come out of their heads!' says Tammy from St Louis, USA. She continues 'since using ear candles my husband has been free of itching ears! He has had itching ears since I met him eight years ago!' Anneliese M from New Brunswick, New Jersey claims that 'through ear candling my hearing has improved – from fifty per cent hearing to about eighty-five per cent'.[59] This resounding figure of 85 per cent is right on the button for the final red flag, which brings, on my count, ear candling's QQ total to thirteen.

Thirteen is an eerily appropriate number since ear candling is certainly unlucky for some. In the only study of its kind to date, a 1996 survey of 144 ear, nose and throat (ENT) specialists revealed twenty-one reports of serious injury caused by ear candling, most being caused by molten wax dripping down inside the hollow candle and directly into the ear. The research, published in the American ENT journal *The Laryngoscope*, described thirteen cases of external burns, seven cases of ear canal obstruction with candle wax and one perforated eardrum.[60] In 1999 Dr Richard Harris of Brigham Young University in the US followed the case of a woman patient who had experienced severe pain and bleeding from the ear following the use of an ear candle. A plate of solidified candle wax had to be removed from the eardrum surface with a surgical hook.[61] There have been at least two house fires associated with ear candling and one Alaskan woman died after her bed caught fire during a solo session. All these cases have led to the import and sale of ear candles for therapeutic purposes being banned in both Canada and the USA, but promoters still try to get round the regulations in both countries by advertising ear candles 'for entertainment purposes only'.[62] [63]

The *Laryngoscope* researchers also tested whether the burning of ear candles could produce any negative pressure in the form of suction, to enable the 'drawing out' of earwax. A measuring device showed none, a good thing since any suction powerful enough to draw out earwax would probably rupture the eardrum. Others have done experiments to see what happens when you light an ear candle on its own, safely away from anybody's ears. Sure enough, after the candle has burned down for a few minutes, a brown residue will be found at the centre of the candle when you cut it open.[64] Laboratory analysis has found this residue to be . . . (drum roll) . . . burnt candle wax. So ear candling turns out to be nothing more than a conjuring trick, but perhaps not one that justifies the claim of being 'entertainment', even in Wolverhampton.

The world's largest manufacturer of Hopi ear candles is Biosun, a company based in Schwalbach, Germany. In the mid 1980s Biosun bought the name 'Hopi' from the Hopi Indian tribe for an undisclosed sum to launch the candle brand. Hopi ear candles are central to Biosun's ambitions in what the company describes as 'the economic mega-trend of the next decade – the future market "Wellness"'. Biosun's Hopi ear candles retail at around £6.50 per pair with your Thermal Auricular Therapist charging upwards of £20 a session. Unlike their Thermal Auricular Therapist clientele, Biosun is carefully non-specific when it comes to making claims. 'Discover Pure Wellness!' they cry, with the 'Indian relaxation ceremony [which] calms the mind and soothes the head and ears'. Biosun never suggests that ear candles can remove earwax. Instead they say that the flame will create 'a vibration of air in the ear candle, generating a massage-like effect on the eardrum' so the user is able to 'relax, let go, revitalise'.[65]

It is thought that the Hopi, a small Native American tribe, were unaware of the fortune about to be made from their brand and they are said to be taking legal advice. The Hopi tribe's own information material makes no mention of ear candling in any of its descriptions of their life, history, religion or culture. Indeed a spokesperson for the Hopi Tribal Council in Kykotsmovi, Arizona, recently told a journalist that ear candling 'is not and has never been a practice conducted by the Hopi tribe or the Hopi people'.[66]

In Britain there are still no Department of Health warnings or restrictions on ear candling and the practice is increasingly popular. It is offered by many high street beauty salons as well as by alternative practitioners. Because of the lack of regulation it is impossible to know how widespread ear candling is, but an internet search of UK websites returns over 488,000 hits. Many parents (mostly mothers – what are they thinking of?) are thought to use ear candles on their children. UK doctors' organisations are so far surprisingly quiet about ear candling, although there are reportedly moves at the principal British Ear, Nose and Throat doctors' organisation to issue a policy statement warning against the practice.

One British ENT specialist, who does not wish to disclose his identity, became so concerned about the ear candling craze that he went undercover to attend a Hopi ear candling course run by a local CAM therapist. He told me he didn't know whether to laugh or cry when he discovered how little the teacher knew about the ear, either its anatomy or what could go wrong with it. After just a seconds-long glimpse in the students' ears with a cheap disposable otoscope she 'candled' each of them as a teaching exercise. The session ended with the ritualistic breaking open of the candles to demonstrate the debris. The teacher appeared only mildly discomfited when every candle showed an identical deposit of the supposed earwax, saying that this was a strange coincidence that had never happened before.

This ENT doctor/spy was intrigued to hear about the Hopi Indians from his teacher, whom she described in her handout as 'the oldest Pueblo people, with great medicinal knowledge and a high degree of spirituality'. He was interested because he already knows something about the Hopi. They are members of a group that enjoys dubious renown in ENT circles. 'If you want to see real ear disease,' he told me, 'go to the North American Indians – they are famous for their terrible ears.'

Chapter 3

Full of Eastern Promise

Since the beginning of the current vogue for alternative medicine, seekers after new remedies have looked East, keen to believe the wildest claims as long as they were ancient or came from an exotic and faraway place – preferably both. This alleged treasure house includes acupuncture and herbal medicine from China and Ayurvedic medicine, yoga, Unani and Siddha from India. But what do we really know about these practices, their history and whether, or how, they work?

Asia offers a multitude of different traditional health practices. Only Traditional Chinese Medicine and Ayurveda have been taken up seriously by Western alternative medicine, so it makes sense to concentrate on them here. But much of what is described shares many basic principles with, or can be generalised to, those other Eastern traditions based on a pre-scientific understanding of the body.

Chinese medicine first hit Western headlines in 1971 when James Reston, a political correspondent for *The New York Times*, developed appendicitis while in Beijing. He was admitted to the nearby Anti-Imperialist Hospital where his appendix was removed under local anaesthetic, but a couple of days later was still suffering painful post-operative cramping. He then received acupuncture and had burning herbs placed next to his abdomen. Reston felt noticeably better within the hour and went on to write a major feature about his experience for his newspaper.[1] Versions of this account were published around the world and the story became so distorted in the repeated telling that many came to believe that the appendectomy itself had been performed under acupuncture anaesthesia alone. It is this tale – a genuinely Chinese whisper – that appears to have triggered the West's current interest in acupuncture and Traditional Chinese Medicine.

The following year saw US President Richard Nixon's visit to China, a trip so diplomatically groundbreaking that composer John Adams was inspired to write an opera about it, *Nixon in China*. Adams describes the visit as 'an epochal event, one whose magnitude is hard to imagine from our present perspective'.[2] The President was accompanied by a team of health officials and doctors who toured Chinese hospitals to see TCM in action. They observed several operations during which patients were apparently given only acupuncture anaesthesia, evidently neither sedated nor hypnotised and afterwards were able to walk away from the operating theatre with no apparent pain or problems. One man, minutes after having a thyroid tumour surgically removed, drank a glass of milk, held aloft a copy of the *Little Red Book* and announced 'Long live Chairman Mao and welcome American doctors!'[3] Initially sceptical, the American medical team was totally won over, with Nixon's own doctor quoted as saying 'I have seen the past, and it works'.[4]

The lifting of the Bamboo Curtain and the reopening of diplomatic relations between the USA and China happened to coincide perfectly with the beginnings of the holistic health movement in the Western world. For alternative medicine it was to be, almost literally, a shot in the arm.

Traditional Chinese Medicine

Countless ancient diagnostic and therapeutic practices have evolved in China over the last five thousand years which collectively have come to be known as Traditional Chinese Medicine. TCM is based on something very similar to the medieval European humoural model, but identifies five, not four, elements related to specific organs and tastes: fire (heart/bitter), earth (spleen/sweet), metal (lung/pungent), water (kidney/salty) and wood (liver/sour).

TCM's big idea is that an invisible life force exists within the body and that all illness is caused by a disruption or interruption of this vital force. This energy, called Qi (pronounced chi) flows through the body along vessels called meridians. But neither these meridians nor Qi have been shown to exist. Acupuncturists themselves are happy to concede that 'you can't look at [Qi] under a microscope, you can't detect it with any scientific instruments. You cannot isolate it in any form or substrate'.[5] Equal and opposite primeval forces, known as yin and yang, are said to control the flow of Qi. Yin and yang have to be balanced in order to enable the five elements to work in harmony, which allows the body to achieve a state of good health. Yang is hot, bright, upward moving, active, exciting and male. Yin is the opposite: cold, dim, heavy, downward moving, passive, inhibited and – you guessed it – female. All TCM therapies are aimed to redress any imbalance (known as bing) in the yin and yang, as well as correcting any overactivity or weakness among the five elements.

The variations of all these components, mixed with other factors like the weather and the time of year, produce a complicated system of diagnosis and therapeutics. Excess yin is thought to manifest itself in problems including diarrhoea and abdominal pain, excess yang presents as a dry heat resulting in a hyperactive metabolism. Wind-cold syndromes include headache, pains in the body and a 'floating' pulse. An attack of damp heat is said to lead to eczema. TCM also holds that an excess of one or more of the 'seven emotions' can cause physical illness: joy, fright, worry, fear, anger, sadness and grief. In the past excess Qi was sometimes attributed to having had sexual intercourse with ghosts, but this diagnosis seems to have been altogether dropped from modern TCM practice.[6] Or perhaps the spirits are less willing nowadays.

This mystical kaleidoscope of cause and effect gives practitioners of TCM enormous power over patients. And we're only just getting to the diagnostic process, which may involve four stages: inspection of the patient's appearance, especially the tongue; auscultation and olfaction (listening to and smelling the patient); taking details of medical history and lifestyle and pulse-taking. TCM holds there are six pulses that correlate with twelve different body organs or functions and these are checked for twenty-five different qualities on each wrist in order to determine which meridians are deficient in Qi. Not surprisingly, since this method of pulse-taking can involve more than three hundred variables, it can take some time.

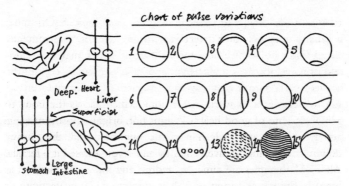

Pulse taking – adapted from the Huang Di Neijing's Suwen (the Yellow Emperor's Canon of Internal Medicine, 2nd Century CE)

The states of different organs are supposedly discerned depending on where and how deeply the physician feels for the pulse.

Chart of Pulse Variations – adapted from the Tuzhu nanjing maijue (Classic of Difficult Issues and Pulse Doctrine, 2nd Century CE)

15 Pulse variations within the Five Phases: 1. Floating (metal); 2. Like a bow-string (wood); 3. Weak (water); 4. Slow (earth); 5. Hollow (fire); 6. Firm (wood); 7. Deep (water); 13. Filled in (fire); 14. Rough (metal); 15. Weak (metal).

The practitioner's aim is to enable the Qi to flow in a balanced and free fashion so as to promote the healthy functioning of the body. TCM's avowed intention is to be a preventative, not curative form of medicine, but this does not stop some practitioners making explicit claims of cure for a range of conditions, including infertility, eczema, high blood pressure, tonsillitis, pneumonia, obesity and dysentery.[7] In recent years Western acupuncture practitioners have become less likely to claim they can manipulate the flow of Qi, tending to restrict their practice to relieving pain, particularly that which is musculo-skeletal in origin, such as osteoarthritis or back trouble. Acupuncture anaesthesia, where the technique is used to provide pain relief during surgery, is the most dramatic example of Traditional Chinese Medicine.

Five thousand or fifty years old?

In many respects the term Traditional Chinese Medicine is misleading. The history of Chinese medicine is as varied as any and superstitious practices have been adopted and abandoned century by century. But its proponents claim that TCM, as described above and available in a high street near you, dates from as long ago as 3000BCE. Those beguiled by all things Eastern believe that both principle and practice have remained constant for millennia. We are encouraged to believe that it has been transposed, intact and un-modified, from ancient times to today. Nothing could be further from the truth.

A number of different schools of medicine developed in China during the last two thousand years, just as they did in Europe. Each had a distinct theory and philosophy of medicine. Although the five-element model is presented as a unified idea today, early subscribers rejected the concept of yin and yang. Until the eighteenth century anatomical dissection was subject to the same taboo as it had been in the West so there was no comprehension of the body's structure or the function of the different organs.[8] There was a belief that there were six yin and six yang organs. The pancreas was not recognised. An undefined fire-related 'triple warmer' organ called the San Jiao occupied the whole of the torso. This still features in modern TCM anatomical descriptions as 'a rather nebulous organ' that can be 'a difficult concept for Westerners'.[9] The spleen (not the brain, which was believed to be a marrow reservoir) was the centre of thought, the liver was where tears were produced and the kidneys were said to be the seat of both willpower and fear.

Early Chinese manuscripts mention nothing resembling acupuncture. Some ancient texts do refer to the use of sharp objects, perhaps made out of flint, for bloodletting or lancing in cases of infection. At this time it was believed that Qi energy travelled through the blood vessels and it is not until the end of the first century BCE that there are descriptions of either twelve, eleven or ten meridians which were said to carry both blood and a life force: Qi or pneuma.[10] The early Chinese had no concept of disease but instead believed illness was caused by evil spirits lodging in the meridians. There are accounts of their disease-causing energy, called 'hsieh', being discharged by making openings with metal lances.[11] The number of suggested openings varies between texts.

Classical acupuncture appears to have involved either bloodletting or cauterisation. Fine needle acupuncture of the type seen today does not appear until the seventeenth century, by which time a total of 365 acupoints had been described, neatly corresponding with the number of days in the year. Today's acupuncture diagrams feature a total of more than two thousand ever-changing needling points and there is no agreement on the number of meridians. Dr Felix Mann, one of the founders of the British Medical Acupuncture Society, has observed that if some modern texts are to be believed, there is no skin left which is not an acupuncture point.[12] And although ancient Chinese medicine had no concept of the continuous circulation of either blood or life forces, modern TCM practitioners now check for circulatory pulses at the wrist in apparent contradiction of the earlier anatomical view.

Dr Paul Unschuld of the University of Munich is the leading Western authority on the history of Chinese medicine. He has said that the origins of Traditional Chinese Medicine as we know it today actually lie in the very recent past, and TCM is 'a misnomer for an artificial system of health care ideas and practices generated between 1950 and 1975 by committees in the People's Republic of China'. Unschuld describes how, after the communist revolution, the vast and heterogeneous Chinese medical heritage was restructured to fit Marxist-Maoist principles. Crucially, TCM was needed to maintain social and political control in a country beset by poverty and with fewer than twenty thousand scientifically trained doctors, mostly practising in big cities, to serve a predominantly rural population of around six hundred million. Ancient practices were selectively cherry-picked, with many elements reinterpreted, in order to 'build a future of meaningful coexistence of modern Western and traditional Chinese ideas and practices'.[13]

The history of the TCM that was exported to the West was reinterpreted to suit Western sensibilities. In an interview published in the journal *Acupuncture Today*, Dr Unschuld says 'it is a fact that more than 95 per cent of all literature published in Western languages on Chinese medicine reflect Western expectations rather than

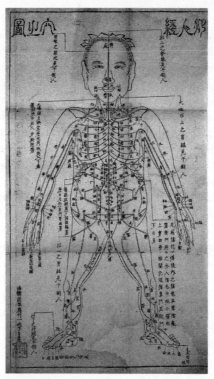

Map of meridians and acupuncture points, Japan, 1700

Chinese historical reality . . . while they reflect Western yearnings, they fail to reflect the historical truth'.[14]

For it turns out that in the early days of Chinese communism there were those, notably the revolutionary Ch'en Tu-hsiu (1879–1942), who ridiculed the ancient medicine and saw it as superstitious, irrational and backward. In an address entitled 'My Solemn Plea to Youth' (1915) he wrote 'Our men of learning do not understand science; thus they make use of yin-yang signs and beliefs in the five elements to confuse the world . . . they know nothing of human anatomy . . . and as for bacterial poisoning and infections they have not even heard of them. We will never comprehend the Qi even if we were to search everywhere in the universe.

All these fanciful notions and irrational beliefs can be corrected at their roots by science, because to explain truth by science we must prove everything with fact.'[15] Many political thinkers shared this attitude before the revolution, with Marxist T'an Chuang describing Traditional Chinese Medicine as the 'collected garbage of several thousand years' in 1941.[16]

But with so few doctors trained in scientific medicine a more influential figure saw the need for a different perspective. Medical historian Kim Taylor's fascinatingly detailed *Chinese Medicine in Early Communist China* describes how Communist ideologues, principally Chairman Mao Zedong, created the TCM we know today. She calls it 'a medicine of revolution'.[17] Taylor describes how, in his 1944 speech 'The United Front in Cultural Work' Mao isolated the enemies of the communist cause as illiteracy, superstition and unhygienic practices. What he went on to say deserves quoting at length because these are the ideas that explain the policies implemented throughout China after 1944:

> In the Shaanxi-Ganse-Ningzia border region the mortality rate of men and livestock is high, and still many of the people believe in witchcraft. In these circumstances, if we only rely on the new medicine we will not be able to solve our problems. Of course the new medicine is superior to the old medicine but if they (the doctors of the new medicine) are not concerned about the sufferings of the people, do not train doctors to serve the people and do not unite with the thousand-odd doctors and veterinarians of the old school in the border region to help them to improve, then they will be actually helping the practitioners of witchcraft by callously observing the death of a large number of men and livestock . . . Our task is to unite with all the old style intellectuals, old style artists and old style doctors who can be used, and to help, educate and remould them.

The message is clear, Kim Taylor insists: 'The "new medicine" was preferable to the "old" but if resources were inadequate, then it was necessary to use the "old".' Straightforward pragmatism would produce a consolidated medicine that could satisfy the nation's health needs and it was Mao himself who enabled the modified version of the traditional medicine to be saved. Certainly his own doctor, in a controversial 1995 biography, describes Mao as rejecting it when ill and saying 'even though I believe we should promote Chinese medicine, I personally do not believe in it. I don't take Chinese medicine'.[18]

Mao's ideas took shape in the slogan 'the scientification of Chinese medicine and the popularisation of Western medicine' which was to dominate Chinese medical policy in subsequent years. In post-revolutionary China *The New Acupuncture* by scientifically trained doctor Zhu Lian became the principal acupuncture manual and placed Maoist propaganda at acupuncture's heart.[19] Though interest in acupuncture had dwindled in the first half of the twentieth century, Zhu Lian believed that it had the right qualities, both practical and political, to serve the Chinese Communist Party. She completed the book in 1949 just as the Communists finally won the civil war and, as Kim Taylor demonstrates, uses military and administrative metaphors throughout. Zhu Lian's acupuncture diagrams show the body in divisions or parts rather than the integrated whole it had been represented as in the past. For the first time acupuncture points are arranged in divisions and straight lines. 'Internal body parts are ascribed a bureaucratic role in the functioning unit of the body, with the heart ascribed the role of ruler, the lung that of ministers, and so on, in hierarchical order.' *The New Acupuncture* is full of these administrative and political metaphors to the extent that 'a direct image of the Chinese Communist Party has been superimposed on the body,' says Taylor.

The 'new acupuncture' was also the means to tackle other public health issues. 'Experience has proven that the masses are accustomed to acupuncture [and] it is of the people, and in this way can successfully further develop mass sanitation movements,' wrote Zhu Lian. The actual therapeutic value was always a secondary consideration. More important politically was that the validation of acupuncture and other ancient practices provided an illusion of health care and that the large number of practitioners of Chinese medicine became a cheap and easily accessible resource.

Suddenly, by acknowledging its 'barefoot doctors' China had no doctor shortage after all. As Kurt Butler describes in his *Consumer's Guide to Alternative Medicine*, in later years both acupuncture and independence became sources of great pride to the Chinese, especially when welcoming Western visitors. 'It became politically correct (and therefore definitely beneficial to one's health) to believe in acupuncture.'[20] The standardised new acupuncture developed in the 1950s was formally disseminated through the textbooks given to the 'barefoot doctors' and other students of TCM. Crucially, these were the acupuncture manuals that were translated into English and became the basis of the acupuncture taught and performed outside China, such as the *Barefoot Doctors Manual* published in America in 1977.[21]

It is worth noting that in China today use of all TCM methods is in decline. A 2006 online survey by the national newspaper *China Youth Daily* and Tencent.com found that 72 per cent would choose Western medicine before TCM. The number of TCM doctors in China has halved since 1949, from 480,000 to 219,000. Further debate about the value of TCM has been stimulated by a petition submitted by Zhang Gongyao, a professor at the Central South University in Changsha, Hunan Province, which urges China's health authorities to remove TCM practices from the national health service. Almost one hundred years after revolutionary Ch'en Tu-hsiu's call to reject the 'ancient medicine', Zhang and his supporters describe TCM as 'unscientific and untrustworthy'; allege that it uses untested concoctions and obscure ingredients to trick patients and then employs a host of excuses if the treatment doesn't work.[22] Those who support TCM say its critics are 'ignoring history'. Chinese health officials have defended TCM by saying it is 'an inseparable and important component of China's health sector' and that 'Chinese medicine has been acknowledged in a growing number of foreign countries'.[23]

Most ironically of all, Mao's vision of the 'Scientification of Chinese Medicine' may still come to fruition through close collaboration with Western capitalism. Since 2000 'Big Pharma', in the form of the Swiss multinational drug giant Novartis, has invested £100 million in a Shanghai pharmaceutical research and development centre. The hope is to isolate any active compounds in Chinese herbs, which can then be developed into prescription drugs by being refined to make them more effective. Paul Herring, head of corporate research at Novartis, explains that 'there are so many compounds in nature, from the seas to the jungles, it's very difficult to know where to start. China has thousands of years' experience of using plants in Chinese traditional medicines. The idea was, why not use the Chinese experience as a kind of filter?'[24]

Other pharmaceutical giants – including Merck, Pfizer, AstraZeneca and Roche are either collaborating with existing Chinese medicine producers or setting up their own Chinese research operations. The new drugs they aim to produce will have a huge domestic as well as worldwide market. Pharmaceutical sales in China alone are worth more than $12 billion a year.[25]

Needle in a haystack – research into acupuncture

Acupuncture has perhaps the highest status of all alternative methods. Even those sceptical about other sorts of alternative medicine are prone to believe that there is probably something in it. They have picked up on a widespread but vague idea about needles doing something unexplained to the body, perhaps triggering the release of painkilling chemicals. And that because there seems to be a kernel of hard science there, the other stuff about Qi energy might be true as well.

As we have seen, the theories on which TCM is based come from a pre-scientific era in which very little was known about the body. Medical science has since demonstrated the heart's role as a pump whose beats correspond to the single rhythm of the pulse, felt at the wrist and elsewhere on the body. TCM practitioners can provide no physiological evidence for the six distinctly separate organ and meridian related pulses they claim to be able to detect. Neither is there is any evidence that TCM tongue examination is a way of diagnosing disease. But even when these mystical mechanisms have been discounted, the belief persists that Chinese medicine can be explored and verified using modern research methods. Over the years thousands of studies have attempted to test whether acupuncture in particular has a measurable effect, making it probably the most researched of all CAM methods. But acupuncture research is a befuddling area. What is the best way to test the effectiveness of acupuncture? And what exactly is it that is being tested?

One of the most effective medical research techniques is to compare the effect of a treatment with a placebo. For obvious reasons acupuncture has presented a particular challenge in this respect, but in recent years researchers at the University of Heidelberg have developed a sham acupuncture needle, designed to be indistinguishable from the real thing. Subjects feel initial pressure from a needle that then telescopes, as if it has penetrated the skin.[26] Research has confirmed that patients cannot tell the difference between sham acupuncture and treatment using real acupuncture needles.[27] This has been a great step forward: subjects can now be effectively 'blinded' so that they don't know whether or not they are having acupuncture. Unfortunately the same cannot yet be done for practitioners, who always know when they are using a fake needle. Ideally they won't know until the moment before the procedure begins, allowing the least possible chance of them communicating, knowingly or not, that information to the subject.

Any acupuncture research also has to answer the question of what is being tested, on a number of levels. Is it 'acupuncture', or simply the effect of needling? Traditional acupuncture, which for convenience let's call the 'Qi' version, holds that needling a specific point on the body will have a specific effect. This can be tricky when different acupuncture methods have conflicting locations of acupuncture points. One Qi version acupuncturist may suggest, for example, that needles placed at points at the ankles will cure sinus problems. Another might suggest points on the scalp. Such disparity led one study to conclude that the most widely used methods of locating acupuncture points 'are grossly imprecise [which] would rule out any specificity or function for acupoints because of the resulting overlap of points'.[28] So not only does any research have to demonstrate a specific effect, but it also needs to test whether random needling elsewhere on the body might also have that *same* specific effect. And even if an effect can be demonstrated, is it achieved by the unblocking and rebalancing of Qi energy or does the physical effect of any old needling provide a physiological distraction – a kind of counter-irritant – from pain or other trouble-some symptoms?

And how can we rely on research findings when geography turns out to have perhaps even more influence than methodology? The findings in acupuncture research vary wildly depending on where in the world it is undertaken. This was starkly demonstrated in a wide-ranging assessment of published research into acupuncture by a group from the UK Research Council for Complementary Medicine in 1998. They found that research conducted in some countries nearly always supported acupuncture. Countries in North America, Western Europe and Australasia reported positive effects in 50 per cent or less of their studies. But 'research conducted in certain countries was uniformly favourable to acupuncture: the results for China, Japan, Russia and Taiwan were 99%, 89%, 97% and 95% respectively.'[29]

Acupuncture anaesthesia has made headlines ever since the publicity following James Reston's Chinese report in 1971 and is surely TCM's most audacious claim. But for years there have been allegations that these tales of acupuncture anaesthesia were either exaggerated or faked. 'Unfortunately it was all a hoax,' alleges Kurt Butler, 'the patients had been carefully selected and indoctrinated, and the demonstrations staged. Many if not all of the patients had been given a tranquilliser, local anaesthetic, and/or painkiller in addition to the acupuncture.'[30] Worse still are the allegations that people were forced to have it for ideological reasons. During the period of the 'Cultural Revolution' of the 1960s and 1970s doctors and patients 'had no choice but to have exceptional courage in order to carry out or undergo surgery [with acupuncture anaesthesia] especially as the patients who felt pain could not cry out. Some resorted to shouting political slogans during surgery in a loud voice,' according to Chinese physicians Keng Hsi-chen and Tao Nai Huang in their 1985 paper on acupuncture anaesthesia.[31]

Neurologist Dr James Taub visited China in 1974 with the Acupuncture Study Group of the Committee on Scholarly Communication with the People's Republic of China. He visited the Acupuncture Research Institute in Beijing as well as traditional medical hospitals in the Shanghai region. The study group was able to 'substantiate a number of previous reports that almost all patients operated upon under "acupuncture anaesthesia" received other additional agents. These almost always included phenobarbital [a sedative] and meperidine [a painkilling narcotic] before and during the operation. Local anaesthesia was also used liberally. I personally witnessed operations in which local anaesthesia was used from beginning to end, but which were nevertheless classified as done under "acupuncture anaesthesia".'[32] Similar practices were noted by a US delegation from the Committee for Scientific Investigation when they visited China twenty years later.[33] A US National Council Against Health Fraud report says that in China today acupuncture anaesthesia 'is not used routinely, but only on the 10% to 15% of people who are suggestible and perhaps easily hypnotisable'.[34] Historian Paul Unschuld reports that in recent years acupuncture anaesthesia in China in major surgical operations has 'slipped into deserved oblivion'. He says that only recently have Chinese doctors been 'able to report, without personal risk, about the pain that patients had been expected to endure through therapy applied in operating theatres not on the basis of scientific knowledge, but in accordance with the ideological precepts of the Communist Party'.[35]

But despite such wide-ranging and long-standing evidence we continue to be persuaded of the seemingly miraculous ability of acupuncture to provide pain relief during surgery. In 2006 an Open University/BBC documentary about alternative medicine featured a heart operation that a Scottish TV reviewer described as showing 'a young female factory worker undergoing open-heart surgery with only acupuncture to control her pain'.[36] But those with a more questioning approach saw something very different. Simon Singh in the *Daily Telegraph* called this memorable bit of television 'emotionally powerful but scientifically meaningless'. He pointed out that 'in addition to acupuncture, the patient had a combination of three very powerful sedatives'. These were midazolam (a powerful sedative used in surgery or investigative procedures), droperidol (a major tranquilliser which was administered as a pre-med in the UK until it was discontinued in 2001) and fentanyl (a heavy-duty narcotic). Large volumes of local anaesthetic were also injected into the chest. 'With such a cocktail of chemicals, the needles were merely cosmetic,' wrote Singh. When he checked with several medics they told him the procedure was 'neither shocking nor impressive'.[37]

Wealth is still a determining factor in Chinese health care. A powerful inducement to have acupuncture might also have been the fact that the operation cost the woman a third of what it would have been under a general anaesthetic.[38] A recent World Health Organisation survey measuring equality of medical treatment placed China 187th out of 191 countries.[39] Another survey by China's own health ministry found that more than 35 per cent of Chinese did not seek medical treatment because they could not afford it. Twenty-eight per cent of those admitted to hospital had to leave because of financial difficulties. One forty-year-old woman with cancer said 'I am having Chinese medicinal treatment because the Western medicine is too expensive . . . it's really too much for me. I might have to stop seeing the doctor next year.'[40] Such considerations are bound to influence people's health choices in any developing country. So when the WHO reports that in Vietnam '50% of the population preferred to be treated by traditional rather than modern medicine,' it is worth bearing in mind that such a preference might primarily have been dictated by the cost.[41] Whichever way you come at it, acupuncture research evidence – of either usage or efficacy – is a challenge to interpret.

American biochemist Thomas J Wheeler concludes in a review of acupuncture research that 'in general, the better designed the study the smaller have been the beneficial effects.'[42] Although some well-designed studies have suggested that acupuncture can be effective in limited sorts of pain relief, such as in cases of knee osteoarthritis, positive results are few and are less likely in randomised controlled trials, where a placebo is used and the researchers don't know who is being treated with what. *Bandolier*, an independent journal about evidence-based health care, reviewed the research evidence for acupuncture effectiveness in treating a range of conditions and found it to be ineffective in treating (amongst other things) asthma, neck and elbow pain, headache, menopausal hot flushes, cocaine addiction and giving up smoking. There is a small amount of evidence that acupuncture can help several kinds of pain, including back pain, some sorts of arthritic and post-operative dental pain. It also seems to reduce post-operative or chemotherapy-related nausea and vomiting in adults, but not in children.[43] [44] But these benefits are tiny in comparison with the fulsome claims invariably made for acupuncture – namely that it can cure specific illnesses as well as act as an alternative to conventional anaesthesia during surgery.

Beyond the cult of Qi

So even though it may be of limited use, by what mechanism could acupuncture reduce pain? One scientific theory suggests that needling stimulates the body to produce its own painkillers, called endorphins and enkephalins, which then flood our system and are able to dull the kinds of pain or nausea described above.

Also posited is the 'gate' theory, which holds that needling might be able to block pain impulses from reaching the brain. Canadian psychologist Ronald Melzack and British physiologist Patrick Wall developed this general theory in the 1960s, and have since discovered that the analgesia (pain relief) produced by acupuncture can be produced by many other types of sensory stimulation, such as electricity or heat both at designated acupuncture points and elsewhere on the body. They concluded that 'the effectiveness of all of these forms of stimulation indicates that acupuncture is not a magical procedure but only one of many ways to produce analgesia by an intense sensory input'.[45] Or as James Reston put it after his Chinese acupuncture experience back in 1971, the insertion of needles 'sent ripples of pain racing through my limbs and, at least, had the effect of diverting my attention from the distress in my stomach'.[46] In this way acupuncture could be acting as a simple counter-irritant which reduces the sensation of other stimuli, like osteoarthritis pain. As the American sceptic and clinical professor emeritus Wallace Sampson says, 'if it has the effect of, say, releasing endorphins, well, many things release endorphins – a walk in the woods, a five-mile run, a pinch on the butt'.[47]

Now Dr George Ulett, a psychiatrist and acupuncturist from

the University of Missouri School of Medicine, has developed an electro-acupuncture technique that he says is an evidence-based and easily learned non-drug method of pain control. Ulett and acupuncturist SongPing Han's book *The Biology of Acupuncture* describes in detail what they call 'the neuro-biological equivalents of such hypothetical constructs as Qi and meridians'. Electrode pads are used instead of needles. When certain 'motor points' on the body, particularly on the wrist, are given electric stimulation, brain chemistry is altered in such a way that a number of chemicals including endorphins are stimulated to circulate around the body and relieve pain.[48] In other words, 'acupuncture acts through the nervous system and not by yin/yang, Qi or meridians'.[49]

Dr Ulett rejects what he calls the cult of Qi. 'It is no longer necessary to spend hundreds of hours learning the magical intricacies of traditional Chinese medicine. Science has made these rituals obsolete.'

Safety of acupuncture

In Britain anyone can set themselves up as an acupuncturist regardless of training or accreditation, with voluntary self-regulation through several professional bodies. Government control of the practice is promised within the next few years but is a long time coming.

With the advent of fine needles the risks of acupuncture have been greatly reduced, especially in the West. In the past there have been injuries such as pneumothorax, where air gets into a lung and causes it to collapse, but these are now rare, especially when the acupuncturist has been properly trained and understands human anatomy.[50] Infection risk has also been greatly reduced with the introduction of disposable needles, but there have nevertheless been cases of hepatitis B infection at the hands of practitioners with poor hygiene practices. Nerve injury, nausea and vomiting, and fainting have all been reported but are again uncommon at the hands of trained and experienced practitioners. Deep needling would not be safe for people who have bleeding disorders or who are taking anticoagulant drugs like warfarin.[51]

Foreign exchange

Beyond mainstream acupuncture there are a number of variations of TCM, all of which have been invented in the last hundred years. Ear acupuncture, for example, is a wholly modern European creation not found in ancient Chinese texts. Often given the more impressive title of 'auricular therapy' it transposes a map of the body on to the ear. The image superimposed is that of an inverted foetus with the lobe representing the head area, the outer edge of the ear the back and legs, the body organs within the central or concha area of the ear. The associated claims are similar to those of whole body acupuncture such as balancing or unblocking the flow of Qi. Specific areas will be needled or massaged depending on the trouble. The intention is to both diagnose and treat and there are unsubstantiated claims that ear acupuncture can cure cocaine addiction and help people give up smoking.

Ear acupuncture was invented in the 1950s by a French doctor working in Lyons called Paul Nogier. The rather implausible story goes that Nogier encountered several people who claimed to have been cured of sciatica after having their ears cauterised by healers (known as *guerisseurs*) in Marseilles. This somehow led him to theorise that the ear was in some way correlated to every other part of the body. It is thought that the Chinese learned of Dr Nogier's work and then developed their own auricular mappings. 'This correspondence system was easy to teach "barefoot doctor" acupuncture technicians to readily assimilate into their paramedical practices,' according to the American journal *Medical Acupuncture*.[52] Dr George Ulett has noted that the vagus nerve, which is linked to all the major body organs, supplies the central section of the ear, but this anatomical fact may be irrelevant since there is little research evidence for the efficacy of ear acupuncture anyway.[53]

Reflexology is another twentieth-century Western creation. Frequently incorporated into TCM in both East and West, it was in fact invented in the 1930s by an American physiotherapist, Eunice Ingham.[54] Reflexology involves superimposing a map of the body on to the feet and the therapist pressing their fingers on the specific points supposed to correlate with other parts of the patient's body. When investigated, diagnoses have once again been found to be both incorrect and inconsistent between practitioners.[55]

Reiki involves the harnessing and transmission of more Qi, prana or universal life energy. Translated from the Japanese, reiki means 'life force'. It was invented – though adherents prefer to say 're-discovered' – in 1914 by Japanese businessman Mikao Usui, following business and marriage difficulties.[56][57] Practitioners claim to be able to transmit healing energy by placing their hands on or near the body. 'You may feel a flow of energy, mild tingling, warmth, coolness or nothing at all.'[58] Some say they can perform reiki at considerable distance. 'Receiving a distance reiki healing energy session is a convenient way to enhance the quality of your life without leaving the comfort of your home,' according to one practitioner, who styles herself a 'Shamballa Multi-dimensional Healing Master'.[59]

Chinese herbal medicine

Chinese herbal medicine is based on the same philosophical principles as acupuncture, with herbs said to enable the free flow of Qi energy, improve yin/yang balance and prevent any elemental deficiency or excess. Chinese herbal medicine has a strong claim to be CAM's oldest therapy, dating back to as long ago as 3000BCE when the 'Divine Farmer' Shen Nong wrote the first known Chinese herbal text.

There are more than five thousand different medicinal plant species in China, but only around five hundred are used with any regularity in the West. Most remedies contain several different herbs and may also contain one of more than three hundred other non-plant substances, derived from animal parts or minerals. These include charred human hair, donkey skin gelatine, squirrel droppings, tiger penis, cockroach and powdered oyster shell – the last said to 'settle the spirit and help yin'.[60] They are sold individually or made up into formulas of as many as twelve different ingredients. Only one or two of the constituents are directed at the major aspects or symptoms of the condition being treated, the others being intended to enhance these effects. The dried herbs are weighed out by the caddy or jin, a measure that has now been standardised

HOW TO PREPARE HERBAL MEDICINE
Put a bag of herbs into a pot, add 1¾ cups or cold water.(1cup=330ml)) boil for 40 minutes, drain the solution out should be about ½ cup left, this is 1st dose. drink it after food today. not really matter when, just once a day. don't throw these herbs away tomorrow re-boil again - but add 1¼ cups of cold water & boil for 20 minutes only. this is 2nd dose. drink it same time tomorrow.

before boil : soak for ½ hours

Chinese herbs prescribed to treat menopausal hot flushes

to 300 grams. This amount is divided further to produce one liang (30 grams), one qian (3 grams) and one li (0.03 grams). A typical combination designed to treat menopausal symptoms may contain six different herbs of varying quantities: one liang of one, half a liang of another, and three qian of each of the others.

The patient takes the herbs home, adds water and boils them down to create a soup or Tang, which is then cooled, filtered and taken on an empty stomach, usually first thing in the morning. 'The herbs will at first taste unusual and often bitter to anyone who has not tried them before, but the vast majority of people get used to the taste very quickly,' according to the Register of Chinese Herbal Medicine, a UK organisation which represents some four hundred practitioners.[61] Alternatively, Chinese herbal remedies are increasing available in powdered, tincture, capsule or tablet form, 'targeted at the more stressed lifestyle of Westerners who have little time to boil extracts in the traditional manner each day,' presumably unlike like those millions who make up the time-rich Chinese leisured classes.[62]

Chinese herbal medicine is Britain's favourite brand of herbalism but also potentially the most dangerous, as we will shortly see. In recent years there has been an explosion in the number of high street herbal medicine shops in the UK, which are now estimated to number more than six thousand. This market is now worth an estimated $200 million and is the biggest in Europe.[63] Chinese herbalists offer treatment for the usual list of chronic conditions treated by alternative medicine, but there are further (all unsubstantiated) claims that these herbs can be used in the treatment of breathing disorders, infertility and hormone disturbances, drug addiction and cancer. In the UK anyone can set themselves up as a herbalist without any qualifications. While the British Register of Chinese Herbal Medicine runs an approved quality-control scheme for its members, which monitors production from farming through to packaging, there is no guaranteed standardisation of either quality or potency in most shops or over the internet.[64] What is more, some of the species' names do not translate accurately from the Chinese, or the same name may be applied to a variety of plant types.

Some Chinese herbs have been found to contain active substances, which have been shown (amongst other things) to have real anti-inflammatory, sedative or antimicrobial effects. Research, albeit on a small scale, has shown that Chinese herbal remedies may be useful in short-term treatment of childhood eczema; however there have been reports of liver damage associated with the use of these and other Chinese herbal treatments.[65] Despite such findings Chinese herbs are still unregulated in the UK and are sold as nutritional supplements, not drugs. When a *Daily Mail* reporter described symptoms of migraine at a total of ten different TCM clinics he was given a dizzying assortment of diagnoses, from 'wind in the brain' to an 'overheated digestive system' and 'blocked liver'. Only one practitioner referred him back to his GP. The rest 'tried to sell me expensive courses of acupuncture and Chinese herbal remedies'. When he asked what one prescription contained he was told it was 'all natural, no side-effects'.[66]

It is almost impossible for consumers to get any hard information about Chinese herbs. Take just one Chinese herbal remedy, dong quai, made of Chinese angelica root. Sold as a food supplement, this is promoted as 'the ultimate, all-purpose woman's tonic herb'.[67] It is said to be able to treat a range of menstrual and menopausal problems, to help manage constipation and treat premature ejaculation in men. No surprise then that such a panacea is a favourite of the developed world's dominant alternative medicine user: the middle-aged, middle-class, educated woman. Dong quai use is perhaps more widespread than other Chinese remedies because it is promoted by CAM practitioners such as Marilyn Glenville, author of *Natural Alternatives to HRT*, some of whom may not have specialist knowledge of Chinese herbs. It is contained in the 'Black Cohosh Plus' capsules she sells on her website.[68] Dong quai can be obtained in various other forms, either as the root itself, or as powder, tablets and gel. There's even a Boots' own brand dong quai product.

The Natural Medicines Comprehensive Database, based in the US, provides objective, scientifically reliable evidence-based data on natural medicines. It has no axe to grind and does not advocate for or against the use of natural medicines. It describes, with full supporting references, what dong quai contains and how it is believed to work. Does this sound like a 'food supplement'? According to the NMCD, dong quai has 'several coumarin constituents including osthol, psoralen and bergapten, and contains 0.4 per cent to 0.7 per cent volatile oil. Some coumarins can act as vasodilators and antispasmodics. Osthol appears to inhibit platelet aggregation and smooth muscle contraction, and cause hypotension [low blood pressure]. Psoralen and bergapten are photosensitising agents and can cause severe photodermatitis. Bergapten and other dong quai constituents, such as safrole and isosafrole, are carcinogenic . . . other research suggests estrogenic effects. Dong quai might stimulate the growth of breast cancer cells.'[69] Furthermore dong quai may interact with anticoagulant herbs or drugs like warfarin, increasing their potency and the consequential risk of bleeding.

And does dong quai work? The database entry concludes that there is no research evidence whatsoever that it is able to reduce menopausal symptoms. There is some evidence that when it is mixed into an ointment including ginseng, clove flower, cinnamon and toad venom it might help to delay ejaculation (though you can't help thinking the last ingredient might have an inhibiting effect all its own). In conclusion it is 'possibly unsafe when used orally and in large amounts. Dong quai contains several constituents that are carcinogenic'. Like any drug, dong quai has risks and (evidently not many) benefits. But while menopausal women will receive detailed information on HRT from their prescribing doctors and on pack information leaflets, they are unlikely to find any such information when they buy dong quai. So even setting aside questions of efficacy, it is simply impossible to make an informed choice about whether to take it, in what form and for how long.

What's your poison?

Other Chinese herbs have been shown to contain dangerous levels of heavy metals such as lead, mercury, arsenic and copper which may either be deliberately added for supposedly therapeutic purposes or occur naturally in the herb. The fact that poisons are naturally present doesn't make them any safer. In some cases contamination may occur during the growing of the herb or during processing.[70]

The stimulant ephedra, for example, is contained in the herb ma huang, often used in Chinese herbal slimming preparations. It has amphetamine-like effects. The American Food and Drug Administration banned ephedra in 2004 following a major study which reported more than sixteen thousand adverse events associated with its use, including heart palpitations, tremors and insomnia. It was also linked to one hundred and fifty-five deaths in the US.[71] There is little evidence that ephedra is effective in boosting physical activities and weight loss, as claimed by its salespeople. But ma huang is still easily available on the internet in products with names like 'Herbalean' or 'Gold Star Nutrition – Original Power Thin'.[72] [73]

Perhaps more alarming are the frequent cases of deliberate adulteration of Chinese herbs with prescription drugs. Research by the California Department of Health Services in 1998 found that when two hundred and sixty Asian patent medicines were investigated 'at least 83 (32%) contained undeclared pharmaceuticals or heavy metals, and 23 had more than one adulterant'.[74] The Food and Drug Administration in America has warned diabetics and people with low blood sugar against using several specific brands of herbal products because they were found illegally to contain the prescription diabetic drugs glyburide and phenformin, and could pose a serious health risk. These have also been found in a product sold as 'Shortclean' in Canada, promoted as being derived from only natural ingredients.[75] Sale of another Chinese herbal mixture, PC-SPES, this time given to people with cancer, was stopped in the US when it was found to be adulterated with the prescription drugs indomethacin (an anti-inflammatory), diesthylstilbestrol (a female hormone also known as DES which has caused birth defects) and warfarin (an anti-coagulant).[76]

A 2002 review of studies of Chinese herbal medicines worldwide found adulteration to be widespread. Suspicion had been raised not only by adverse effects but also by unexpectedly good ones. Added pharmaceuticals have included the sedative diazepam (Valium), anti-inflammatories, steroids, anti-convulsants and sildenafil, commonly known as Viagra. Overall the proportion of Chinese herbal medicines likely to contain synthetic drugs is unclear and

An inadvertent truth in a Chinese herbalist's window

varies from country to country depending on what controls exist, but that it is a serious problem is beyond doubt.[77]

These international findings are consistent with UK reports. In 2003 researchers from Sheffield tested twenty-four herbal creams advertised as 'natural' and intended for use in the treatment of skin conditions including childhood eczema. These are creams bought over the counter from herbalists, health food shops and by mail order, mostly by mothers who want to avoid giving prescription preparations to their children. Twenty were found to contain 'powerful or very powerful' corticosteroid drugs. Steroid creams can improve eczema dramatically but are not prescribed for children because they can cause permanent skin damage, retard growth and disrupt hormones. One of the products, called Wau Wa, appeared to be the cream Dermovate (which contains a potent corticosteroid) mixed into a paraffin base. Dermovate should not be used on the face, long term or by children, but Wau Wa's instructions were for it to be applied 'all over' with no advice on for how long it should be used or the recommended minimum age for treatment.[78]

In 2006 BBC Radio 5 Live revealed sixty-seven ongoing investigations by the UK government's Medicines and Healthcare products Regulatory Agency (MHRA) into the selling of illegal medicines by Chinese medicine stores. Most of these involved adulteration with pharmaceuticals. At the Herb Garden store in Leigh-on-Sea, Essex, a reporter was sold a herbal slimming pill and told it was made from rhubarb and honeysuckle. Laboratory tests showed it contained fenfluarmine, a drug banned in most countries because it is dangerous, including the UK. The reporter was also sold preparations containing two other prescription-only drugs: danthron, a powerful laxative usually prescribed only to terminally ill patients because it is carcinogenic and sibutramine, a drug prescribed only in cases of extreme obesity and which can cause a dangerous rise in blood pressure.

The BBC report quoted one patient, David Woods, who had visited the Herb Garden store for acupuncture on his painful knees. The practitioner told him he 'should lose a bit of weight and it would help my knees. She said she had these new pills and would I like some?' The pills turned out to contain fenfluarmine that has caused permanent damage to his heart. 'I honestly thought I was dying,' he said. Dr Karl Metcalfe, a consultant physician at the local hospital, reported treating nine other patients who had taken the pills but feared there could be many more. 'For a medically qualified person to be issuing these drugs would be reprehensible. For a non-medically qualified person to be doing it is alarming and quite clearly criminal,' he said.[79]

Because of this lack of control over Chinese herbalists it is impossible to know exactly how many other people might have been similarly poisoned. In the UK between 2000 and 2004 there were four hundred and five suspected cases of adverse reactions to herbs (not all of them Chinese) reported to the MHRA. A third of these were serious, such as liver problems (ranging from abnormal liver enzymes to hepatitis), herb–drug interactions, blood clotting problems and allergic reactions. Remedies made from the herb aristolochia were linked to two cases of kidney failure in the UK in 1999. Elsewhere, one hundred Belgian women developed irreversible kidney damage after being given aristolochia in error at slimming clinics. Thirty needed kidney transplants, others dialysis.[80] [81] But as is so often with such products, attempts to ban the substance in the UK and elsewhere have been undermined by its availability on the internet. Similarly, the psychoactive root kava kava was banned in many European countries following seventy cases of liver problems worldwide, including four deaths and seven liver transplants, but remains readily available by mail order.

Representatives of the Register of Chinese Herbal Medicines may decry the dangerous activities of high street herb suppliers, but they appear to have little authority over the worst kind of practitioner. They are powerless to stop, for example, the sale of the herb Fufang Lu Hui Jiaonang, used to treat constipation and heartburn, which has been found to contain mercury at levels 117,000 times above the legal limit for food in the UK.[82] In 2006 the MHRA confirmed that recent samples of Chinese herbs found in the UK have contained mercury, toxic herbal ingredients (often where a toxic herb has a similar name or appearance to the intended ingredient), potent prescription-only medicines and human placenta. But despite the wealth of evidence of adulteration, contamination and inadequate guidelines for use, the MHRA is unable to offer the consumer much advice beyond this generalised warning: 'The public should be aware that there are some TCM products on the UK market that may be manufactured to low quality standards and may be deliberately adulterated or accidentally contaminated with toxic or illegal ingredients. These products do pose a direct risk to public health and it is not currently possible to distinguish between these products and TCMs that are made to acceptable safety and quality standards.'[83]

Ayurveda and the Goblet of Metabolic Fire

We move now from ancient Chinese wisdom to ancient Indian wisdom, in the form of Ayurveda, the dominant form of traditional medicine in India. 'Ayur' means life and 'veda' means knowledge or science. It must surely win the prize for being the most ancient form of traditional medicine given that an Indian government department says it 'originated with the origin of universe'.[84] In India there are more than 360,000 Ayurvedic practitioners and more than two thousand specialist hospitals. Since the early 1970s there have been moves in India to give Ayurveda a more scientific and evidence-based research foundation, but such research is still to be published and a science-based approach is likely to be rejected elsewhere. This view expressed on the 'AyurBalance' website is typical of the Western alternativist approach: 'Veda means science: not a science that changes its theories and its findings every few years but ageless, eternal knowledge built on siddhantas, fundamental unchanging principles'.[85]

Ayurveda was introduced to the West via the Maharishi Mahesh Yogi, of Transcendental Meditation fame and one-time Beatles' guru. Like Traditional Chinese Medicine, Ayurveda is founded on a humoural model of five elements: ether, air, fire, water and earth. It sees the human body as consisting of three doshas or bio-humours (vata, pitta and kapha), dhatus (the material of the body) and malas (excretable products like faeces, urine and sweat). The metabolism is governed by agni or metabolic fire, which has thirteen different forms and pulse diagnosis and a general physical examination are used to detect disease and imbalance. Treatment is meant to address both mind and body and may involve a mixture of some of the hundreds of Ayurvedic medicinal herbs – up to sixty at a time – which can be given in a variety of ways: orally, oil drips, massages or heat treatments. Ayurvedic medicine is also full of non-contentious and sensible information about the need for a healthy lifestyle, advising moderation in work, diet and exercise, eating healthy food, a regular daily routine, daily meditation and taking time to enjoy the simple things in life like nice sunsets and spending time with your family.

Illnesses or imbalances are treated with Ayurvedic herbs. The spice turmeric is used to treat rheumatoid arthritis, wounds and Alzheimer's disease; guggul, a tropical shrub resin, is used to treat obesity and reduce cholesterol; a mixture of sulphur, iron, powdered dried fruits and tree root treats liver problems.[86] As with TCM, many Indian herbs have been shown to contain pharmacologically active substances that are possibly useful, but research evidence is limited. However, also in common with Chinese herbal medicine, there have been many cases of poisonous ingredients and adulteration. One American study looked at seventy different Ayurvedic products on sale in the Boston area and found that fourteen (20 per cent) contained lead, mercury, or arsenic. Some of these were in very high concentrations, hundreds or thousands of times greater than the maximum recommended levels.[87] This risk has been confirmed in a British report that showed adulteration of Indian medicine with heavy metals is a problem worldwide.[88]

Deepak Chopra is Ayurveda's most prominent advocate in the US and Europe. A trained endocrinologist, Chopra worked for the Maharishi before setting up on his own. Chopra is also a founder member of the Alliance for A New Humanity, an anti-materialist organisation which aims to end war, poverty, environmental degradation and human rights abuses by creating 'an alliance of people based on the awareness of humanity's interconnectedness.'[89] His books sell in their millions and include *Quantum Healing, Ageless Body, Timeless Mind, The Seven Spiritual Laws of Success* and *Golf for Enlightenment*. In his best selling *Perfect Health* he introduces Ayurveda, which tells us 'that freedom from sickness depends on contacting our own awareness, bringing it into balance, and extending that balance to our body'. You will also learn 'the power of natural healing to transcend the ordinary limitations of disease and aging'.[90]

Deepak Chopra's HQ is the Chopra Center for Wellbeing in La Jolla, California. He is associated with a range of training courses and retreats, which run throughout the year in various locations, usually at expensive spa resorts. You might start with the 'Perfect Health Program' created by Chopra and his colleague Dr David Simon, which has 'the intention of soothing you, teaching you, and immersing you in practical behaviours that will allow you to return to your world with renewed vitality, radiant beauty, and a reconnection to your most healthy state'. This costs $2,750 for five days and includes a daily Ayurvedic massage, a single thirty-minute lifestyle consultation, a Perfect Health workbook 'signed by Deepak' but neither transport, food nor accommodation. If, however, you want to become an accredited teacher of Deepak Chopra's Ayurvedic approach you will have to pay thousands of dollars and complete a course made up of a prescribed series of workshops: Journey into Healing ($1,475), Seduction of the Spirit ($1,775), Primordial Sounds Meditation ($325) and a Home Study Program ($1,795). Once again, accommodation is not included but for lunch you can 'try delicious Chopra Center Ayurvedic recipes each day for only $15 per day'. If money is tight you can still take the free 'What's your dosha?' quiz on Chopra's website. Using this I was able to discover that my nature is 'Vata' which means I have a 'buy, buy, buy' shopping style and could maximise health and wellbeing by drinking Deepak Chopra Relaxing Tea and enrolling in one of his 'Perfect Health Programs'.[91]

If you live in the UK you will never have to travel far for a bit of Ayurvedic attention, but it's maybe not quite so flash. There's the 'Vedic retreat' health farm in Burton-on-Trent where you can 'loose weight naturally' (*sic*) or the Tor Spa Retreat in Canterbury, both of which significantly undercut Chopra's rates.[92] At Tor Spa, for an astonishing £50, you can have a 'Body Prayer massage' and 'feel Prana (life energy) circulate and de-stress you, with this deep thorough meditative massage, designed to open the nadis (breath channels) and harmonise the seven chakras. Allow one hour'.

The champions of both Traditional Chinese Medicine and

Ayurveda have been quick to understand the essential appeal of their brands and have made the most of the great susceptibility in the West to the 'mystic East'.[93] The success of TCM and Ayurvedic medicine arguably may be testimony to the entrepreneurial skills of their practitioners, but it doesn't mean that they are either safe or effective.

Chapter 4

Eurotrash: herbs, homeopathy and the mystery of the missing molecules

In August 2005 a headline flashed around the world. It was 'The end of homeopathy' according to a huge study published in *The Lancet*, which found no difference in effect between homeopathy and placebo.[1] But it wasn't the end, not by an infinitely long chalk. The *Lancet* report may have concluded that homeopathy was no more potent than a dummy drug, but its editorial was still forced to lament 'the fact that this debate continues, despite 150 years of unfavourable findings. The more dilute the evidence for homeopathy becomes, the greater seems its popularity'.[2]

Nothing, it seems, can halt the worldwide progress of what sceptics see as Europe's biggest CAM scam. Homeopathic medicine is based on the idea that diseases can be cured by substances that produce the same symptoms as the disease; the Greek word *homios* means similar, *pathos* means suffering. These substances are diluted many hundreds of times before being administered, to the point – which both supporters and detractors agree – where there is no medicine in the medicine. But the greater the dilution, asserts the homeopath, the greater the benefit.

Homeopathy is often the first alternative remedy that people try, making it the gateway drug of the complementary medicine habit. Around a quarter of Europe's population uses homeopathy. Eighty-five per cent of Belgian GPs and nearly half of all family doctors in Holland prescribe it. It is increasingly popular in Latin America, Russia, India and other Far Eastern countries.[3] It is used by at least six million Americans.[4] Whilst homeopathy's lack of proven effectiveness may have led the Swiss government to decide to stop paying for it under their health insurance system, usage in Switzerland is undiminished.[5]

In the UK there are five government–funded homeopathic hospitals and an estimated 42 per cent of GPs either prescribe it themselves or will refer patients to homeopaths.[6] The royal seal of approval is bestowed by Queen Elizabeth II, and homeopathy is said to be used by her extended family and pets. When she travels abroad she is understood to be accompanied by a leather case packed with sixty vials of homeopathic remedies.[7] The British monarchy's own website describes how European royalty has used homeopathy for years: 'Sir John Weir, a royal physician, is reported to have prescribed homeopathic remedies for three Kings and four Queens attending the funeral of King George V in 1936. King George VI, also a subscriber to homeopathy, even named one of his racehorses Hypericum' after a homeopathic remedy for depression.[8] Prince Charles is known to be a keen user and the Queen is patron of the Royal London Homeopathic Hospital, recently refurbished at a cost to the state of £20 million. Homeopathy is, in every sense, well established in Britain.

Homeopathy has not always enjoyed such good fortune. In the two hundred years since it was invented it has drifted in and out of fashion. As we have seen, homeopathy in the USA nearly died out in the early twentieth century following the regulation of medical schools and the growing success of scientific medicine. In Europe it continued to thrive, but has a chequered history. It was popular in Nazi Germany, where a three hundred-bed modern homeopathic hospital was built in Stuttgart in 1940.[9] Karl Kötschau, a leading spokesman of 'organic medicine' in Nazi Germany, hoped for 'an end to curative medicine, in that homeopathy as part of the new German Science of Healing as part of the National Socialist Revolution would render it unnecessary, to be replaced by preventive care.'[10] Professor Edzard Ernst describes how in the 1930s 'leading German scientists . . . were charged with testing homeopathy at a basic science level in addition to thorough clinical research. The results, which survived the war but later seem to have disappeared in the hands of homeopaths, were apparently wholly negative'.[11]

A more disturbing twist, described by UK homeopath Francis Treuherz, saw homeopathy being used as a weapon of the Final Solution. 'It was said that Rudolf Steiner [the Hungarian founder of the Anthroposophic spiritual freedom movement] used a high potency of the ashes of the spleen, testes and skin of a rabbit, sprayed as a liquid, on to an agricultural estate in Koberwitz, Germany in 1924, to rid the estate of rabbit infestation', writes Treuherz. 'There is a sinister sequel, namely the experiments with potentised ashes of the same parts of young Jews . . . sprayed across the length and breadth of the Reich. This appears to be well documented and the perpetrators have been executed for crimes against humanity.' Treuherz explains this abuse by saying that 'the evidence shows that the Nazis were using whatever form of medicine suited their purposes, whether organic or orthodox at different times'. Homeopathy understandably suffered from such associations after World War Two and they are an aspect of its history rarely mentioned today.[12]

Despite the setbacks that homeopathy has faced, both by association with such users and at the hands of its many legions of scientific detractors, its believers remain undeterred and are convinced of its efficacy. Within months of *The Lancet*'s scathing editorial, the BBC was reporting that 'a new study is boost to homeopathy': a six-year study at the Bristol Homeopathic Hospital, involving more than 6,500 patients, had shown that more than 70 per cent of those with chronic diseases reported positive health changes after homeopathic treatment. Of the group, 75 per cent felt 'better' or 'much better,' as did 68 per cent of eczema patients under sixteen. The greatest effect was seen in children, with 89 per cent of those with asthma reporting improvement. Dr David Spence, clinical director and consultant physician at the Bristol Homeopathic Hospital and chair of the British Homeopathic Association, was a co-author of the study. He told the BBC 'these results clearly demonstrate the value of homeopathy in the NHS'.[13]

So with such an overwhelming number of satisfied users, could there be something in it? You decide.

Less is more – the story of homeopathy

When Samuel Hahnemann became a physician in Germany in the late eighteenth century, medicine was still based on the model of the four humours: blood, black bile, yellow bile and phlegm. Doctors tried to restore humoural balance by a number of brutal and distressing practices that all too often did more harm than good, including bloodletting, cupping, leeching, purging or giving medicines made from mercury or arsenic. Hahnemann rejected the idea of the four humours and instead, inspired by his work as a translator of ancient texts into German, developed a belief that disease was caused by a disruption of the body's *vis vitalis*, or vital force. Illness amounted to what he called 'dynamic disharmonies of our existence and nature'. He also rejected the traumatic treatments of his era and sought a more gentle form of medicine.

In common with most physicians before the development of germ theory, Hahnemann believed that illnesses like smallpox or

Samuel Hahnemann, inventor of homeopathy

measles were transmitted via a 'miasm', a force similar to magnetism. He described contamination taking place invisibly, at distance 'with no more transmission of any material particle from one to the other than from the magnet to the steel needle. A specific, spirit-like influence communicates smallpox or measles to the child nearby, just as the magnet communicates magnetic force to the needle'.[14] Hahnemann believed nearly all chronic conditions were brought about by miasms, the most prevalent of which he described as *Psora*, the manifestation of a suppressed itch. 'This Psora is the sole true and fundamental cause that produces all the other countless forms of disease, which, under the names of nervous debility, hysteria, hypochondriasis, insanity, melancholy, idiocy, madness, epilepsy, and spasms of all kinds, softening of the bones, or rickets, scoliosis and cyphosis, caries, cancer, fungus haematodes, gout, yellow jaundice and cyanosis, dropsy, gastralgia, epistaxis, haemoptysis, asthma and suppuration of the lungs, megrim [migraine], deafness, cataract and amaurosis, paralysis, loss of sense, pains of every kind, etc, appear in our pathology as so many peculiar, distinct, and independent diseases.' The other two principal miasms in Hahnemannian theory are syphilis and sycosis (gonorrhoea). These ideas are still supported by modern homeopaths. According to Dr Manesh Bhatia, leading homeopath and editor of the internet journal 'Homeopathy 4 Everyone', Psora is 'the real fundamental cause and producer of innumerable forms of disease. It is the mother of all diseases and at least seven-eighths of all the chronic maladies spring from it while the remaining eighth spring from syphilis and sycosis. Cure is only possible by proper anti-miasmatic treatment,' by which he means homeopathy.[15]

The central homeopathic doctrine of *simila similibus curentur* (like cures like) or what Hahnemann named the 'Law of Similars' appears to have occurred to him in 1790, when he was forty-four. A similar idea had been suggested by the ancient Greek Hippocrates and by Paracelsus in medieval times. And the 'Doctrine of Signatures', in which illnesses are supposedly cured by plants or substances that resemble them – the white spotted leaves of lungwort for chest conditions for example – is another philosophical precursor of homeopathy.

In the late eighteenth century cinchona bark, which contains quinine, was the recommended remedy for malaria. Hahnemann wanted to understand how it worked and noticed that taking cinchona bark produced malaria-like symptoms in a healthy person. He went on to explore this effect by taking the remedy himself, experiencing amongst other things heart irregularities, fever and thirst – the well-known symptoms of malaria. Modern research has suggested that Hahnemann might have been allergic to quinine as his detailed reports describe taking only four grams of cinchona powder, containing only a tiny amount of the ingredient he was investigating. His reaction was highly unusual, leading one sceptic to say 'it can be concluded that Hahnemann suffered from hyper-sensitivity to quinine. If so, this means the fundamental doctrine of homeopathy is based on a pathological condition of its founder'.[16]

Be that as it may, Hahnemann used the Law of Similars to go on to make a massive generalisation following his single experiment with cinchona: that disease symptoms could be eradicated from the body by a substance which causes those same symptoms in a healthy person. He believed that this process stimulated the body's own ability to heal itself. This was in stark contrast with the prevailing view that illness symptoms should be brutally suppressed or counteracted. Samuel Hahnemann rejected that idea and called it allopathy – from the Greek *allos*, meaning other or opposite.

Hahnemann, together with a group of followers, went on to experiment with the idea of homeopathy by taking a series of natural substances – often poisons like arsenic and strychnine – and recording their effects in detail. He called these experiments provings, and provings remain the basis of homeopathic remedies today. A healthy person takes each of the various substances over a variety of time periods, from days to months, with every single subsequent physical, mental and emotional event and experience carefully recorded, as they happen, minute by minute. One recent account of a proving of *argentum sulphuricum*, for example, includes the report that 'crashing pots in café area don't put me off' and 'as I get ready for work my nose looks bigger'. Physical symptoms include 'teeth hurt after eating soup' and 'feel hot but no fever and feel as if sweat is trying to come out but no sweating'.[17]

Such detailed accounts are frequently contradictory, with a single substance often described as inducing both a symptom and its exact opposite. So the proving of the remedy *veratrum album* may describe the production of both diarrhoea and constipation, *sulphur* both excessive appetite and no desire to eat, *silicea* a mind that is sometimes yielding and sometimes obstinate, and so on. These contrasting reactions are explained by modern homeopath George Vithoulkas as 'a characteristic manifestation of the action of the defence mechanism and therefore must be accorded equal importance'.[18] Sceptic Jay Shelton, in his book *Homeopathy – How it Really Works* does not accept this rationale: 'That a remedy can induce or cure opposite symptoms is so remarkable that one is inclined to consider other explanations. Over the months of a proving it would not be surprising for some provers to experience opposite symptoms naturally: feeling energetic and tired, having diarrhoea and constipation, feeling happy and sad, having a dry and runny nose. No remedy is required to produce these symptoms'.[19]

Since Hahnemann's time other more controversial (even amongst homeopaths) proving methods have been invented. One such is 'meditation proving', where the substance is not taken orally but the provers meditate together about the substance and share their experiences. Another is 'dream proving' in which a substance is taken by a group at the same time and their dreams over the following nights are noted.[20] These proving methods are associated with the creation of new remedies based on more abstract substances, known as 'imponderables', which do not need to be based on an existing material. Recently created remedies have contained sunlight and electricity (both somehow 'captured' in alcohol) as well as those harnessing the healing powers of 'thunderstorm' and 'Berlin Wall'. The latter is said to be helpful in treating monomania, an inordinate or obsessive zeal for a single idea.[21] [22]

The remedies that appear in the modern homeopathic *Materia Medica* (the reference book of homeopathic therapies and their effects) are all accompanied by their proving date, with many listed as being proved in the early nineteenth century. There are more than two thousand remedies in total, but only five hundred or so are in regular use. Remedies may be derived from plants, animals, chemicals or minerals. Provings show belladonna, for example, to cause delirium and hallucinations, so Hahnemann's Law of Similars dictates it can be the appropriate treatment for a condition with those symptoms.

At first, given the toxic nature of many of Hahnemann's 'like cures like' remedies, his research collaborators and patients were frequently troubled by serious and unpleasant side effects. This led him to experiment further by diluting the medicine. Not surprisingly, the more he diluted the fewer the disagreeable effects and the happier everyone was. And as well as suffering fewer treatment-related problems quite a few of the patients got better, as ill people often do. But this eventuality led Hahnemann to the conclusion that the very act of dilution had increased the therapeutic effect of the medicine. Less must be very decidedly more, an idea he expressed as 'the Law of Infinitesimals'.

He went on to develop a procedure whereby an extract of a natural substance, known as the mother tincture, is sequentially diluted at a ratio of one part medicine to ten parts water. He then would place this mixture into a container and shake it vigorously or rap it one hundred times against a hard but elastic object, usually a leather-bound book. Hahnemann believed this process, which he called succussion or dynamisation, released dynamic forces from the diluents, which are preserved and intensified with subsequent dilutions. Hahnemann's *Organon of Medicine*, first published in 1810, describes what he thought was happening: 'This remarkable alteration in the properties of natural bodies through mechanical action on their smallest particles of trituration and succussion . . . develops the latent dynamic powers which were previously unnoticeable, as if slumbering. The dynamic powers of these substances mainly have an influence on the life principle, on the condition of animal life. Therefore this process is called dynamisation or potentisation (development of medicinal power). Its products are called dynamisations or potencies of different degrees.'[23] This is what modern homeopaths still believe.

This dilution and succussion process is repeated a second time, resulting in a dilution ratio of 1:100. The next, third, dilution gives 1:1000, a fourth 1:10000 and so on. Each dilution adds another

'100 strong succussions with the hand on a hard but elastic body'
amounts to Hahnemann's 'first degree of dynamisation'

nought and all the while, according to Hahnemann, the forces associated with the mother tincture are getting stronger rather than
weaker. Indeed, in line with Hahnemann's writings, the degree to
which the substance is diluted is rather confusingly described in
homeopathy as its potency. There are two main ways of describing
the series of dilutions. A tenfold dilution is denoted 'x' and a centesimal (hundredfold) dilution is denoted 'c'.

In the previous paragraph we left our substance languishing at
a fourth dilution, which would be expressed as 4x by a homeopath.
Back in the real world there is a limit to any possible dilution that
can be made, after which the remedy has to all intents and purposes
disappeared and the odds are against any sample of the solvent
(in this case the water) containing a single molecule of the solute
(the mother tincture).

So, to the mathematics of homeopathy. A 'mole' is the molecular weight of a substance, expressed in grams. There is a large
but finite and specific number of atoms or molecules in a mole.
This number of atoms or molecules is 6.022×10^{23}, also known
as Avogadro's number, after the eighteenth-century Italian

mathematician and physicist whose work led to the formula. When Avogadro's number is applied to homeopathy it means that after the 24x – twenty-four dilutions at 1:10 parts – dilution there is virtually no chance of a single molecule of the original remedy being present in any sample taken from the dilution. This is known as the dilution limit and means the remedy has been diluted to the extent that a single molecule of it can appear in only one dose of the many hundreds or thousands prepared. The rest are just water.

In homeopathic preparations 30x, however, is a common dilution level – way beyond the dilution limit. 30x means the remedy has been sequentially diluted one part in ten a total of thirty times. Each dilution adds another zero, so the final dilution is one part medicinal substance to 1,000,000,000,000,000,000,000,000,000,000 parts water. A single dose of homeopathic treatment is taken from that mixture and dropped on to a small pillule made of lactose or sucrose, described as 'a neutral medium', where it evaporates.[24]

It's hard to get your head round all these zeros. To understand the scale of dilution that's involved it helps to see what would happen if we had added water to the original volume of mother

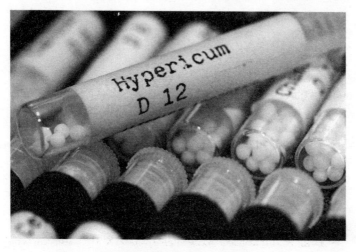

Homeopathic 'sucrose pillules'

tincture. Dr Stephen Barrett of Quackwatch calculates it by suggesting we imagine a cubic centimetre of mother tincture – that's one fifth of a teaspoonful – mixed with enough water to produce a 30x dilution. This volume of water is greater than the amount needed to 'fill a container more than fifty times the size of the earth. Imagine placing a drop of red dye into such a container so that it disperses evenly. Homeopathy's "Law of Infinitesimals" is the equivalent of saying that any drop of water subsequently removed from that container will possess an essence of redness'. Physicist Robert Park provides another way of conceptualising the 30x dilution. He calculates that to be precise, at a dilution of 30x you would have to drink 7,874 gallons of the solution to be guaranteed one molecule of the medicine. Or if you prefer you could eat at least one thousand tons of lactose tablets in order to be sure of getting the same single molecule of the homeopathic remedy.[25]

Homeopathically speaking, 30x isn't even much of a dilution. Since it is believed that the greater the dilution the greater the potency, many remedies are diluted by the hundredfold (centesimal or c value). 'If the problem is mild in nature' says a practical guide to homeopathy, a 6c potency 'is an appropriate dose to start with'. If the problem 'is more intense and severe in nature, has developed slowly and insidiously over three or four days, and is making the patient emotional, anxious or upset, a 30c is deemed more appropriate'.[26]

As for what these dilutions mean in volumes one can readily understand. I came across a member of a homeopathic discussion internet list, who gives his name only as Hans, who has helpfully worked it all out on a spreadsheet. This shows homeopathic ratios in various units of measure. Remember that at 1c the medicine is at one part in 100? The next dilution adds two more zeros, so at 2c the medicine is present at one part in 10,000. By the time the medicine is at just 6c it has been diluted a million million times, which Hans describes as the equivalent to one drop of remedy in four Olympic-sized swimming pools. Another seven dilutions give you 13c, one drop of remedy in three quarters of the earth's oceans. 16c is one drop in a sphere the size of the earth, 19c is one drop in ten spheres with the same diameter as our solar system and 26c is one drop of remedy in a sphere with the same diameter as the Milky Way. The typical 30c homeopathic dilution is equal to one drop of mother tincture in nearly a hundred million galaxy-sized balls of water. 'Perhaps this explains why sceptics are saying there is no active substance in the remedies,' concludes Hans, with admirable understatement.[27]

Putting it another way, 30c is the equivalent of one molecule of remedy dissolved in 1,000,000,000,000,000,000,000,000,000,000, 000,000,000,000,000,000,000,000,000,000 molecules of water. That is a lot of water. It would fill a container more than 30,000,000,000 times the size of the earth. That is a very big container.

The $20 million dollar duck

Oscillococcinum is a popular homeopathic remedy for flu, derived from the heart and liver of the Barbary duck and created by a French doctor, Joseph Roy, during the influenza epidemic of 1917. Looking through his microscope at the bodily fluids and tissues of flu victims Roy observed some unusual-looking bacteria, which consisted of two unequal balls he called oscillococci because they appeared to be jumping. He began to see these micro organisms everywhere – in cancer tumours, syphilitic ulcers, the tubercles of tuberculosis patients and in the pus of gonorrhoea sufferers. He also detected what he was by now calling the 'universal germ' of oscillococcinum in people who had eczema, rheumatism, mumps, chickenpox and measles.

Dutch sceptic and mathematician Jan Willem Nienhuys takes up the story: 'It is not clear today what Roy saw through the eyepiece of his microscope. But one thing is certain: he did not see the causes of those diseases. Rheumatism, eczema, and most forms of cancer are not caused by microbes, and mumps and measles are caused by viruses, which can't be seen with an ordinary microscope. Moreover, no other bacteriologist has ever reported seeing Roy's special cocci again . . . Roy's finding fitted perfectly with the homeopathic view that diseases do not have specific causes, and he thought that his discovery could be adapted to treat cancer homeopathically'. No one knows why Roy decided to choose the Barbary duck as his source of oscillococci. It is thought he might have used the liver because it was believed to be the seat of suffering in ancient times. Nienhuys suspects that perhaps the French tendency to call any form of illness a 'crise de foie' or liver crisis might also have had something to do with it.[28]

The standard dilution of homeopathic oscillococcinum is a zero-popping 200c. I'd better let Professor Park do the sums again. 'That would result in a dilution of one molecule of the extract to every 10^{400} molecules of water – that is 1 followed by 400 zeros. But there are only about 10^{80} (1 followed by 80 zeros) atoms in the entire universe. A dilution of 200c would go far, far beyond the dilution limit of the entire visible universe.'[29]

In 1997 an American magazine noted that this degree of dilution meant only one bird a year need be sacrificed in order to manufacture the worldwide supply of oscillococcinum pills. 'In a monetary sense, this single French duck may be the most valuable animal on the planet' wrote reporter Dan McGraw, since the flu remedy generates sales worth more than \$20 million a year for Boiron, its French manufacturer. 'For duck parts, that easily beats out foie gras in terms of return on investment.'[30]

A delicate constitution

As well as consulting the list of remedies in the homeopathic *Materia Medica* to see what fits the symptoms, a homeopath will assess the nature or 'constitution' of the patient and match the remedy accordingly. 'The constitution of an individual is more or less fixed and the remedy which stimulates the recuperative ability will usually stay the same throughout life', according to one handbook published by Oxford University Press. In homeopathy everyone has a constitutional type that matches up with a remedy and this is ascertained by the clinician during an early consultation. So 'one talks, for example, about a *Nux vomica* type, a *Sulphur* or *Pulsatilla* type. These types arise empirically during provings, where it is noticed that certain kinds of constitutions react strongly to certain remedies.'[31]

Here, from three different sources, is how these three particular 'types' might be described:

'The typical *Nux Vomica* patient is well described by the "type A personality". They are impatient, competitive and ambitious. They are irritable and impatient and particularly hate waiting in lines or in traffic. They also tend to be highly sensitive to light, noise, odours and other stimulants.'[32]

'The essence of a *Sulphur* type: among our great leaders, inventors, scientists, gurus, and mathematicians, there have been many *Sulphurs*. The scholar, the creative genius, the eccentric, the impulsive artist; all have minds that flow profusely into metaphysics, theories, and equations.'[33]

'*Pulsatilla* female weeps easily. Timid, irresolute. Seeks open air; always feels better there, even though she is chilly. Wants to hold the head high. Is uncomfortable with only one pillow. Fears in the evening of being alone, ghosts. Likes sympathy. Morbid dread of the opposite sex. Religious, melancholy. Highly emotional, mentally like an April day. Dislikes butter.'[34]

These same remedy types will also be prescribed for certain acute conditions, so for example, *nux vomica* (diluted strychnine) may be given for hangovers, headaches, a tickly cough, or 'extreme over-sensitivity, anger and irritability. Digestive complaints include indigestion, vomiting, diarrhoea, cramps and constipation'.[35] *Sulphur* is prescribed for itchy skin conditions, urinary tract infections or when you are 'touchy, irritable and self absorbed with a tendency to philosophise about the smallest issues'.[36] *Pulsatilla* (derived from the windflower or meadow anemone) is often given for measles. That's the least of it, as it is also recommended for acne, anxiety, asthma, benign prostatic hyperplasia, bronchitis, chicken pox, common cold, conjunctivitis, cough, diarrhoea, dysmenorrhoea, ear infections, haemorrhoids, indigestion, measles, menopause symptoms, menstrual problems and PMS, morning sickness, mumps, oedema, osteoarthritis, post-partum depression, pregnancy and birth support, rheumatoid arthritis, varicose veins and yeast infections.[37]

A homeopathic remedy may also be chosen on the basis of visual similarity to the affected part, 'Doctrine of Signatures' style. 'Thus *Sepia* (squid ink) corresponds to the female reproductive organs.'[38]

For better for worse

Hahnemann died in 1846 but his *Organon of Medicine* remains homeopathy's principal text. Since then there have been occasional adjustments and additions to the homeopathic credo. In the nineteenth century the German physician Constantine Hering gave his name to what is known as the Law of Cure (so many Laws, so little time!) based on his idea that when the patient was being cured then the symptoms moved from the vital internal organs to the less vital organs and finally to the skin, and that they also disappear downwards – head to toe – in the body and in the reverse order of their development. Hering also believed that the body seeks to externalise disease and that symptoms will surface as part of the curative process. This is expressed in homeopathy as an aggravation or healing crisis, so 'if symptoms move outwards, for example, if skin eruptions occur in an asthma patient, a curative process has been stimulated'.[39] Hering emigrated to the USA in the 1830s, where he is known as the father of homeopathy.

This notion of aggravations enables homeopaths to have it every which way. If you get better, all well and good, but if you get worse you are actually getting better. This increases the chance that 'the natural course of disease may be interpreted as a positive result and allowing negative results in clinical trials to be rationalised,' as biochemist Thomas J Wheeler puts it.[40] Thinking is confused and confusing on this point, with some homeopaths seemingly unsure whether these healing crises should be coming or going. UK homeopath Beth MacEoin, in *Homeopathy – the Practical Guide for the 21st Century*, writes 'sometimes there can be a brief intensification of symptoms if you give too high a potency of the most appropriate remedy. This is a good sign, confirming your choice of remedy, but you should stop taking the remedy in order to prevent prolonged aggravation of your symptoms'.[41]

Homeopathic themes and variations

There are various methods of homeopathic prescribing. In *classical* homeopathy a single remedy is prescribed according to the individual's presentation or history. In modern *complex* homeopathy more than one remedy will be used simultaneously. In *fixed* prescriptions the same single remedy is used for a group of patients, an approach often used in research trials when a remedy is compared with a placebo in a large group of subjects. In *isopathy* a homeopathic preparation is based on what is thought to be the illness's causal agent. This is what homeopaths call a nosode. A nosode could be anything from pollen for hay fever to cancerous tissue for cancer. In *phytotherapy* herbs are prescribed, usually in low potencies.

'Tissue salts' are another popular form of homeopathy. They are made from one of twelve inorganic compounds such as 'nat. mur.' or natrium muriatricum, also known as sodium chloride. That's common salt again. I'm detecting a pattern here. Dr Wilhelm Heinrich Schuessler, born in Germany in 1821, who thought that health problems could be caused by a deficiency or imbalance of essential minerals, created this so-called biochemic system of medicine towards the end of the nineteenth century. Tissue salts are usually taken in a low 6x potency and by now you'll have grasped the idea that 'low' is how homeopaths would describe a higher concentration of medicine in the medicine. Nat. mur may be given in the event of anything from 'weight gain with cravings for salty, savoury and fatty foods' to being 'withdrawn and depressed with an inability to cry' as well as 'high blood pressure' and 'head cold with nasal discharge that runs like a tap and then gets blocked'.[42]

Homeopathic prescribing and manufacture was revised further by James Tyler Kent at the beginning of the twentieth century. An American, he was largely responsible for the adoption of the higher centesimal homeopathic potencies in Britain. He helped introduce 'centesimal fluxion machines' which contained rotating glass phials that could be filled, succussed and emptied repeatedly over many hours, without human assistance, starting with a drop of tincture and alcohol or water solutions. High dilutions had already been pioneered by Russian aristocrat General Count Iseman von Korsakoff in the middle of the nineteenth century. Fluxion machines provided a very convenient, mechanised way of making high potencies in a short space of time – in hours rather than weeks. These days homeopathic preparations are made in factories using fully automated industrial-scale potentising machines. The French company Boiron, which dominates the European market, is listed on the stock market and has yearly sales worth more than 360 million euros.[43]

In the UK the economic health of the homeopathy industry can only improve following a move by the Medicines and Healthcare products Regulatory Agency, which must have delighted homeopaths. The MHRA has changed its rules so that manufacturers can claim health benefits for homoeopathic remedies without providing proof of clinical efficacy. So it is now possible to buy a homeopathic product called Coldenza which manufacturer Nelsons say is 'specifically designed to bring fast, effective relief for the symptoms of cold and flu'. Coldenza contains a 6c potency of gelsemium sempervirens, 6c being way beyond the dilution limit and, as you might remember, the equivalent of one drop of remedy in four Olympic-sized swimming pools. Other products in this new range include 'Candida', 'Noctura', 'Polenna', 'Rheumatica', 'Sootha', 'Travella', and 'Teetha'.[44]

The British Pharmacological Society had serious concerns about this development because there is no convincing scientific evidence that homeopathic remedies work any better than placebo, and warned that the MHRA endorsement of the use of such remedies 'may put patients at risk of delayed diagnosis'.[45] The development was also raised in the House of Lords, where Lord Taverne complained 'under this new regulation, the sole basis on which claims of efficacy can be made for homeopathic products quite legally is "homeopathic provings". There is no need for clinical or scientific tests. Homeopathy is not based on science and is not a science in any sense'.[46]

Shaken vacc

Supporters of homeopathy sometimes claim that it has a philosophical link with vaccination, in which minute quantities of a substance stimulate the immune system to protect the body against infection. In reality the fields of immunology and homeopathy are very different. Unlike most homeopathic medicines, vaccinations do actually contain something other than water – a measurable volume of molecules of dead or weakened organisms which incidentally become less, not more potent at a greater dilution. Called antigens, these are related to the known causative agents of the diseases they prevent, such as the smallpox vaccine that is made from live vaccinia, another 'pox' type virus. Those antigens are what stimulate the production of measurable antibodies that protect the body if it encounters the dangerous smallpox virus. Crucially, vaccines are designed to prevent diseases, not to treat them or provide symptom relief as homeopathy claims to be able to do.

Homeopaths anger scientists, therefore, by allying their practices with vaccination. Injury is added to insult if homeopaths go on to advise parents not to have their children vaccinated, offering, as many do, high dilution and ineffective homeopathic nosodes instead.[47] And in 2006 the organisation Sense about Science discovered that homeopathic preparations were being sold in the UK as a means to protect against malaria, putting international travellers at considerable risk. Their researcher had ten separate homeopathic consultations during which she said she was travelling to parts of the world where there was a known risk of malaria. On every occasion she was 'advised to use homeopathic products instead of being referred to a GP or conventional travel clinics where effective medicines are available'.[48]

The consultation

In a leaflet written for the UK's National Eczema Society, the Queen's homeopath Dr Peter Fisher describes a typical homeopathic consultation. 'The homeopath needs to know the details of the eczema. But he or she also takes into account other health problems and more general characteristics: the physical type, psychological make-up and so forth. The idea is not so much to cure the disease as to restore the body's balance, so that the body deals with the problem itself. We always try to see the individual, his illness and life-situation as a whole. Homeopaths talk of "pictures", meaning the overall pattern of the person and illness ... this is quite different from the conventional medical approach.'[49]

The initial consultation is the longest, lasting for an hour or more, and is what appears to provide homeopathy's distinctive sense of purpose and satisfaction for both doctor and patient. Dr Trevor Thompson is a homeopath who is also a GP and lecturer in primary care at the University of Bristol Medical School. He told me that homeopathy 'is more satisfying in a way [than orthodox medicine] because homeopathy leads to a fuller type of case history . . . the emotional element is surprisingly important.'

This positive experience is confirmed in *Passionate Medicine*, a collection of essays written by five doctors and two vets who moved from conventional medicine to homeopathy. The contributors all seem to have been profoundly disturbed by their early experiences of anatomical dissection in medical school and are generally critical of their training 'in particular the notion that effective clinicians should be objective and emotionally detached'. Instead they argue 'for a more holistic, caring model that regards self-knowledge, passion and the ability to create successful practitioner–patient relationships as central to the healing process'. One homeopathic doctor writes 'the more I explored the meaning of illness, the more I found patients would tell me their whole story. It was and is surprising how ready people are to reveal themselves. It also surprised me how therapeutic patients found this, and how grateful they were'. This last, perhaps unintentionally revealing remark highlights what must be a significant difference between the doctor's experience of ordinary medical practice and the homeopathic consultation.[50]

What is beyond doubt is that homeopathy is popular with users – both practitioners and patients. They like it because there are no unpleasant side effects and they truly believe it works. This experience of homeopathy 'working' happens repeatedly. As Dr Trevor Thompson puts it, explaining his continued belief in homeopathy, 'it has to come through results because you would soon get fed up with it if you worked all day and didn't see results'. He believes his greater knowledge of the patient is crucial to a consultation's outcome. 'The paradox of homeopathy is that the more you understand about homeopathy, the richer the map you have of the human territory and the more you understand the predicament of the patient, and therefore the more accurately you prescribe the medication. My experience is that the closer the match between what I perceive to be the patient's reality and my knowledge of the *Materia Medica* the better the results I get.'[51]

Dr Thompson's brand of certainty and positive commitment to

healing may be universal amongst homeopaths but problems arise when you look for evidence beyond what the *Guardian's* Dr Ben Goldacre tartly described as the 'customer satisfaction survey' of the Bristol observational study, in which a group of people who expressly asked to use homeopathy told their doctors they felt better afterwards.[52] The trouble is that clinical trial after clinical trial shows no benefit of homeopathy over placebo, which is where we started with the 2005 *Lancet* meta-analysis. In 2000 the NHS Centre for Reviews and Dissemination looked at systematic reviews of more than two hundred randomised controlled trials of homeopathy and concluded there was 'insufficient evidence of effectiveness to recommend homeopathy as a treatment for any specific condition'.[53] Other studies showing lack of effectiveness of homeopathy include ones looking at pain and bruising after hand surgery, rheumatism, earache, asthma and migraine.[54]

In response, homeopaths say placebo-controlled double blind testing is not a good way to test homeopathy because the remedies are not administered in an authentic individualised way, where two patients showing similar symptoms may receive different treatments. This argument does not wash with two leading complementary medicine researchers, Klaus Linde and Dieter Melchart. They say 'if the claims of classic homeopaths – that striking responses to homeopathic treatment occur quite frequently, that overall homeopathy is a clearly effective therapy, and that the remedy is the main cause of induced changes – are correct, it should be much easier to obtain more convincing and consistent results, even under the problematic conditions of double-blind placebo-controlled trials'.[55] But it is not as if homeopaths themselves feel under any obligation to prove the value of their remedies. After doing some research showing that placebos got better results than homeopathy in the treatment of rheumatoid arthritis, Dr Peter Fisher concluded 'we have come to believe that conventional RCTs [randomised controlled trials] are unlikely to capture the possible benefits of homeopathy . . . It seems more important to define if homeopathists can genuinely control patients' symptoms and less relevant to have concerns about whether this is due to a "genuine" effect or to influencing the placebo response'.[56]

This is where an exploration of homeopathy starts to resemble wrestling with a giant jelly. Despite all the evidence showing that homeopathy *doesn't* work homeopaths keep on trying to demonstrate *how* it works. A notable example came in 1988 when the French biologist and immunologist Jacques Benveniste said he had completed an experiment which showed that a laboratory allergy test worked even when the testing substance was diluted well beyond the dilution limit. The research was published in the UK science journal *Nature*, incensing the scientific community. An attempt to replicate the experiment, attended by three independent observers including US sceptic and magician James Randi and filmed for the BBC 2 *Horizon* programme, failed miserably.[57] Since then further efforts have been made by homeopaths to support Benveniste's observations with varying degrees of success. But as physicist Jay Shelton says, for something as revolutionary as this, much more independent evidence and, ideally, a theoretical explanation would be needed for the ideas to gain acceptance.[58] Towards the end of his life – he died in 2004 – Jacques Benveniste compared himself with Galileo and claimed homeopathic remedies could be transmitted over the telephone and the internet.[59]

It becomes even harder to find homeopathy credible when Dr Trevor Thompson asserts that 'poetry is very similar to homeopathy because how does a poem work? A poem works by being a close match between the word on the paper and some sort of other reality'. He goes on to tell me 'when you meet with someone and you choose a medicine for them, the medicine in some way becomes invested with some of the properties that were established during the time when the practitioner and the patient got together, so they are not separable in the normal pharmaceutical sense . . . there is some sort of wider picture here and homeopathy is tapping into it and it's just even the names of homeopathic medicines, you know like compared to drug names, they're just so much more beautiful.'

All this talk of poetry and beauty and pictures and metaphysics must be quite soothing when there is nothing seriously wrong with you, or if you've got something from which you will recover regardless. But, to quote the journalist Simon Hoggart, have you noticed that homeopathic hospitals never have a casualty ward? 'Try a one in a million tincture of arnica when you get a broken leg,' he says.[60]

Misty water-coloured memories

Homeopaths readily concede that at greater dilutions there are no remaining molecules of medicine in their remedies. Instead they say that homeopathy is explained by the idea of 'the memory of water', the premise Jacques Benveniste thought he was demonstrating. The Queen's homeopath Dr Peter Fisher calls it 'information medicine' and described it thus to *New Scientist* magazine: 'I think it's to do with changes in the structure of the water . . . somehow information is stored in water'. Dr Fisher suggests that the act of preparing a homeopathic remedy imprints this information on the water, adding 'the reason why people find [homeopathy] so challenging is that we are used to thinking about pharmacology in molecular terms.'[61]

At this point homeopaths often turn to quantum theory in search of an explanation both for their experience of homeopathy 'working' and for a theoretical model that would explain their notion of the memory of water. Concepts such as 'non-locality theory' and the 'entanglement model' are borrowed from theoretical physics to support Hahnemann's Laws of Infinitesimals and of Similars. Physicist Jay Shelton understands how homeopaths might find these quantum mechanical theories and explanations attractive but he notes that most of those doing the speculating know little about physics and that their speculations often demonstrate an incorrect application of the concepts. Such theories are also 'fatally flawed because they are basically inapplicable to large and/or un-isolated systems such as cells, people and pills'.[62]

Any exploration of homeopathy invariably leads to a series of questions that no homeopath is able to answer, well before the moment we go quantum. Since a homeopathic remedy is made by placing a highly diluted homeopathic remedy on to a sugar pill from which it then evaporates, shouldn't homeopaths have to demonstrate that sugar has a memory too? Does the memory linger on in the water-based cells of the human body while the pill is digested and excreted, and then get carried away by the sewerage system to the water treatment works and beyond? And what about all the other substances the homeopathic remedy and the water molecules may have encountered in the past? Most of the water on earth has been around for over four billion years; there's been a lot to remember. How and why would water discriminate between remembering a homeopathic 'mother tincture' and remembering, say, the soap used in Marilyn Monroe's bathwater or the saltiness of a prehistoric sea? Can water have a selective memory?

The myth of allopathy

Perhaps Samuel Hahnemann's most pernicious legacy, however, is his invention of the word 'allopathy', which today has become synonymous with orthodox or conventional medicine. Its use is now so widespread that many scientifically trained doctors are quite happy to use it to describe themselves. Hahnemann derived it from the ancient Greek words *allos* (other or opposite) and *pathos* (suffering or illness) to mean a medicine that involves the treating of symptoms with treatments that suppress or oppose them.

Perhaps this may have been accurate in Hahnemann's day, when his fellow physicians believed that illness was caused by humoural imbalance and corrected by the prescription of 'opposites', frequently harsh and distressing methods like bloodletting or purging. But such rationales are long gone in a modern scientific medicine nowadays more concerned with treating an underlying cause – be it bacterial, viral, genetic or environmental – rather than simply suppressing symptoms.

The contemporary representation of medical doctors as 'allopaths' who don't address the causes of disease is nothing more than clever rhetoric, suggests the American public health expert William T Jarvis. He thinks it is a way of creating the impression that the difference between alternative and scientific medicine is a matter only of conflicting philosophies, rather than that of a vitalistic ideology versus science.[63] In the past the term allopathy was considered offensive by mainstream medicine: the word was used as 'a form of verbal warfare . . . "allopathy" and "allopathic" were liberally employed as pejoratives by all irregular physicians of the nineteenth century, and the terms were considered highly offensive by those at whom they were directed'.[64]

Certainly in the early twentieth century the term allopathy was still vehemently opposed and resented by medical doctors. Jarvis quotes a 1902 book for new medical graduates to show just how objectionable it was. 'Remember that the term "Allopath" is a false nickname not chosen by regular physicians at all, but cunningly coined, and put in wicked use against us, in his venomous crusade against Regular Medicine by its enemy, Hahnemann, and ever since applied to us by our enemies with all the insinuations and derisive use the term affords. "Allopathy" applied to regular medicine is both untrue and offensive and is no more accepted by us than the term "Heretics" is accepted by the Protestants, or "Niggers" by the Blacks.'[65]

And now, in the twenty-first century, the US National Council against Health Fraud suggests 'the duplicity of the term aids those who wish to misrepresent medicine' and recommends that it should not be used in reference to either standard medicine or medical doctors. But the 'false nickname' has become so entrenched in the language that this advice may be too late.

Flower power

Dr Edward Bach (rhymes with match) created his flower remedies in the 1930s. After being trained in orthodox medicine, he incorporated homeopathy into his private practice in Harley Street where he also conducted some immunological research. His first wife died and a second marriage ended in divorce. Soon afterwards he made the dramatic decision to abandon his London home and career, destroying all his research papers before moving to the village of Brightwell-cum-Sotwell in Oxfordshire. The Bach Centre, still based in Bach's home, Mount Vernon, describes how he became 'determined to devote the rest of his life to the new system of medicine that he was sure could be found in nature. Instead [of pursuing scientific medicine] he chose to rely on his natural gifts as a healer, and use his intuition to guide him. One by one he found the remedies he wanted, each aimed at a particular mental state or emotion . . . He found that when he treated the personalities and feelings of his patients their unhappiness and physical distress would be alleviated as the natural healing potential in their bodies was unblocked and allowed to work once more'.[66]

It was in the lanes and fields around his home that he found the thirty-eight remedies that make up the Bach system. 'He would suffer the emotional state that he needed to cure and then try various plants and flowers until he found the one single plant that could help him. In this way, through great personal suffering and sacrifice, he completed his life's work.' Spare a thought for Dr Bach, experiencing 'mental torture behind a cheerful face' before going for a short walk and finding he could put himself right with agrimony. Imagine how hard it must have been to get himself into the lanes of Oxfordshire until he discovered hornbeam to be the answer to 'procrastination, tiredness at the thought of doing something'.

The Bach Centre describes how this English country garden version of the more complex Teutonic homeopathy is realised. 'The sun method involves floating flower heads in a clear glass bowl filled with natural spring water. This is left in bright sunlight for

The doorplate at Mount Vernon, home of Edward Bach

three hours, then the flower heads are removed and the energised water is mixed half and half with brandy. The boiling method involves putting flowering twigs into a pan of spring water and boiling them for half an hour. The pan is then left to cool, the plant matter removed, and again the water is mixed half and half with brandy.' In both cases the resulting mix is the mother tincture, which is then diluted to 5x with more water. A tiny drop of this water is put into a 10ml bottle, which is finally topped up with brandy – described on the label as 'grape alcohol' – and sold for around £3.99. That's watered-down brandy at £399 per litre.

Described as 'every woman's emotional ally' the best-selling treatment Rescue™ Remedy is brandy containing a tiny drop of a mixed recipe of five different remedies (cherry plum, clematis, impatiens, rock rose and star of Bethlehem, at 5x dilution in water) which together 'help deal with any emergency or stressful event. Taking a driving test, exam nerves, speaking in public, after an accident or an argument . . . In an emergency Rescue™ Remedy can be taken neat from the bottle, four drops at a time, and as frequently as required. Otherwise put four drops in a glass of water and take frequent sips until the emotions have calmed'. In a smart combination of the alternativist traditions of both East and West Rescue™ Remedy is advertised as 'Yoga in a Bottle'. And look how they've trademarked the word 'rescue'.

Another British manufacturer of Bach products has invented a 'gentle "no taste" alcohol-free formula', i.e. water-only, version of Rescue™ Remedy. Ainsworth's Emergency Spray 'although originally created to cope with acute anxiety and crisis situations (bereavement, separation, loss etc), can be taken by everyone, including children, pregnant mums and pets'. It contains an added ingredient 'which specifically deals with all fears and any resultant negative thought patterns'. 21mls of Emergency Spray costs £6.36. That's a tiny phial of 'no taste' water sold to a willing public for the equivalent of more than £300 a litre.[67] Ainsworth's R&D department is also ahead of the game with 'Clearing Essences' which you spray round a room 'to clear and purify both mind and body while naturally cleansing the space'. An initial burst of 'Cleanse and Protect' clears 'atmosphere in places where disagreement has been expressed'. (This one is useful in the car apparently, which certainly rings a bell with me.) Afterwards you can choose between 'Inner Guidance' for meditation and calming hyperactive children, 'Calm and Strengthen' for confidence and solidity or 'Ground and Go' for a happy stable atmosphere and reawakened joy for living. These room sprays cost just £9.99 for 110 ml.

If you are moved by this to become a Bach flower remedy therapist the training sounds reassuringly undemanding. Philip M Chancellor's *Handbook of the Bach Flower Remedies* describes how to proceed. 'Always remember that the artful practitioner or physician is a good listener! Cultivate this habit and let the patient talk, but be sure to listen attentively! We say this, because in telling us about his physical symptoms, the patient will reveal a great deal about himself, and that is the information that we practitioners are after. He might tell you, all unwittingly, that he is *afraid* the complaint will worsen (mimulus) or that he has *lost hope* of ever becoming cured (gorse). He might say, "I get so *impatient* and or so *tense* that my work is affected (impatiens)."'[68] Training is also offered at the Bach Centre in Oxfordshire, where the prescribed series of three courses – a correspondence course, six days at the centre and six months writing up the case studies of work with real patients – will cost a total of around £1,000. Once you have completed these you are allowed to join the Bach Foundation's International Register and use its logo, although in practice anyone can set themselves up as a Bach therapist without any formal schooling in the method.[69]

There is no published research evidence to support the use of Bach remedies, nor is there any forthcoming. A representative of the Bach Centre says 'we have never set up experiments, and we don't document the help we give to people. When Dr Bach entrusted his work to [his inheritors], and in so doing set up the Bach Centre, he instructed them to keep their lives simple and their work with the remedies simple as well. We don't see it as our role to "prove" that the remedies work – instead we simply demonstrate how to use them and let people prove the effect on themselves.'[70]

The last word has to go to John Diamond, writing in his final book *Snake Oil and Other Preoccupations*: 'When we read about how Dr Bach discovered his flower remedies it makes perfect sense on a sort of flowers-are-harbingers-of-good level which wouldn't have grasped the public imagination quite so forcefully, I imagine, if he'd used thirty-eight types of spider to produce the Bach spider remedies.'[71]

European herbal medicine

Plants provided humankind's first medicines and many plants contain biologically active compounds. Many modern drugs have herbal or plant origins. Aspirin was derived from willow bark, morphine from the opium poppy, and digoxin from the foxglove. The cancer drugs Taxotere and vincristine came from the yew tree and the periwinkle respectively; there are anti-malarial drugs derived from the shrub artemesia.

Research, together with refinement of manufacture, has enabled these plant-based substances to be produced in a purified form, designed to deliver specific and controlled amounts of an active ingredient. Nowadays this is often synthesised with the aim of creating a substance that is safer, more predictable and effective than the plant version. The drug digoxin, for example, may be derived from the digitalis found in foxglove leaves, but the leaves themselves are almost never used today. But while conventional medicine may sometimes employ substances derived from plants, herbal medicine itself is something very different and is based on an entirely different philosophy from those prescription drugs available in an orthodox pharmacy. The *British Medical Journal*'s 'ABC of Complementary Medicine' identifies three major differences between herbal and conventional medicine.

- *Herbal medicine uses whole plants rather than purified extracts.* The use of a whole plant is promoted as an advantage, enabling it to function 'synergistically'. The American alternative medicine guru Andrew Weil, for example, is reported as saying that 'in the case of drug plants, the whole forms, being complex mixtures and therefore impure, tend to be safer than their unmixed derivatives, freed from diluents and made available in a highly refined form.' But the medicinal plant expert Professor Varro E Tyler, who died in 2001, dismissed Weil's idea by pointing out that 'synergism occasionally occurs, but for every case where a desirable action is enhanced, there are several where undesirable actions are produced'.[72] It is also not possible to know the exact quantity of the active substance in the product and the amount and quality may vary depending on growing conditions and other variables in production.

- *In herbal medicine two or more herbs may be taken together.* This common practice of combining herbs means it may be difficult to know which, if any, is the therapeutic component.
- *Herbal practitioners often use non-orthodox diagnostic methods* such as vega testing or naturopathy. Or they might suggest, with no supporting evidence, 'that arthritis results from an accumulation of metabolic waste products' and prescribe diuretic or laxative herbs alongside herbs with supposed anti-inflammatory properties.[73]

Around £126 million a year is spent in the UK on herbal medicines and the World Health Organisation has estimated the global market to be worth £41 billion.[74] Herbs are often prescribed for mild or temporary conditions but some are promoted as cures for serious illnesses like cancer or HIV/AIDS. Herbal medicine is available mostly in the form of loose herbs, tablets, capsules and extracts. But plant substances such as ginseng, gingko or guarana are increasingly available as ingredients of 'functional foods' or 'nutraceuticals' in the form of snacks, teas or soft drinks.

Even though herbal medicine may be very different from conventional pharmacotherapy as prescribed by your orthodox doctor, the fact remains, as the UK government's Medicines and Healthcare products Regulatory Agency (MHRA) reminds us, that 'herbal remedies are genuine medicines which can have a significant effect on the body. Like any medicines, they have the potential for interactions and side-effects'.[75] Herbal medicine carries more risks than any other form of complementary medicine because of its potentially toxic nature. Comfrey, for example, has been linked to cases of irreversible liver damage (with deaths reported); pennyroyal oil with kidney, liver and nerve damage; lobelia with breathing difficulties, rapid heartbeat and low blood pressure.[76]

It is worth restating that the fact that herbal remedies are natural doesn't make them safe, however appealing the idea. That 'natural' has come to be synonymous with 'good' shows just how out of touch we are with the natural world. In fact, to quote modern herbalist Varro E Tyler, 'some plant constituents are among the most toxic substances known. Acutely toxic alkaloids, ranging alphabetically from aconitine to zygadenine, are abundant. Other constituents, such as the peptide amatoxins in certain fungi, can also kill. Even so, consumers are probably less likely to suffer from acute strychnine poisoning by eating Nux Vomica seeds than they are to receive exposure from milder and less obvious toxins through repeated use of such remedies as sassafras or comfrey'.[77] Professor Edzard Ernst of the University of Exeter suggests that under-reporting may mean what is currently known about the adverse effects of herbal remedies may well be the 'tip of the iceberg'.[78] The fact that herbs are so frequently self-prescribed means that such effects are unlikely to be recorded or investigated by, for example, the MHRA in the UK or the Food and Drugs Administration in America.

The risk of herbs being adulterated with pharmaceuticals and other substances has also been well established, especially in relation to herbs from the Far East. This does happen but appears less widespread with herbs of European origin, especially since the introduction of the Traditional Herbal Medicines Registration Scheme in the UK in 2005. Products registered under the scheme will need to meet specific standards of safety and quality – but not effectiveness – and be accompanied by information leaflets.

This Europe-wide system is aimed to protect consumers, but is opposed in the UK by an odd alliance of alternative supplement manufacturers and retailers working with the Conservative Party and celebrities including Jenny Seagrove and Elton John. The organisation calls itself 'Consumers for Health Choice' and objects to the scheme on the basis that it will 'force many popular products that are defined as food supplements into the same regulatory regime as pharmaceutical drugs, threatening their continued availability'. They have, however, plenty of time to adjust to the new rules – products don't have to be registered until 2011. In America meanwhile there are around two hundred and fifty herbs on the Food and Drug Administration's 'Generally Recognized As Safe' list but some of these are culinary, not medicinal herbs. All the herbs are regulated as dietary supplements despite many of them being prescribed as medicines, so once again do not need to have proved themselves either safe or effective.

Government regulations, regardless of country, will unfortunately offer no protection for consumers buying remedies on the internet, which has become a profitable marketplace for unscrupulous herb retailers who blur the line between providing information and product marketing. This is confirmed by Harvard medical researchers who evaluated the claims made on 443 websites providing information on some of the most popular herbal supplements, like gingko biloba, St John's Wort, echinacea, ginseng, garlic, saw palmetto, kava kava and valerian. They found that 76 per cent of such sites were either selling the product or directly linked to a vendor, 81 per cent made health claims, with 55 per cent claiming to treat, prevent, diagnose or cure specific diseases. The researchers concluded that consumers may be misled by vendors' claims and that more effective regulation is required.[79]

Another study, from the University of Texas, surveyed 150 web sites dealing with three popular medicinal herbs: ginseng, gingko and St John's Wort. Thirty-eight sites contained statements that could lead to direct physical harm if acted upon, including false suggestions that a particular herb could protect against disease and that patients should self-medicate instead of seeing a doctor; one hundred and forty-five had omitted vitally important information, such as the risks associated with St John's Wort used in combination with prescription drugs.[80] When herbs of doubtful efficacy and potential risk are then sold in an unregulated and under-informed environment there is therefore bound to be a compounding of risk.

More worryingly, herbal products are often inadequately labelled. Recommended doses may vary wildly between manufacturers and the amount of active ingredient in the product may bear little resemblance to what is promised on the pack. Surveys by the Good Housekeeping Institute and the *Los Angeles Times* in the US found this to be a frequent problem with St John's Wort products, confirmed in 2004 when tests on St John's Wort tablets done by Taiwanese researchers found huge variations in how much active ingredient five different brands contained.[81] The amounts fell far short of the dose claimed on the labels. One tablet contained only 1.7 per cent and the best only contained 38.5 per cent of the stated dose.[82]

Fortunately there are some ways of finding reliable information on the internet. The Natural Medicines Comprehensive Database, described earlier, lists more than one thousand herbs and dietary supplements, and is available on yearly subscription and updated daily. ConsumerLab.com also publishes independent research into vitamins, minerals, and herbal supplements.[83] New York's Memorial Sloan-Kettering Cancer Center's herbal/botanical database carries useful evidence-based information about alternative cancer treatments.[84] There are no equivalent resources in the UK.

Herbal risks and benefits

Though some have been in use for thousands of years, the characteristics of most herbs are still not fully understood. However, what is known is that two herbs in widespread use, St John's Wort and echinacea, may interact with as many as one in four pharmaceutical drugs. American pharmacologist Dr Christopher Gorski explains that substances contained in these herbs can cause the body to metabolise drugs either too slowly or too quickly, either increasing drug toxicity or causing a reduction in a drug's therapeutic effect. He has discovered that the two herbs increase activity of a specific enzyme – cytochrome P450 3A4, since you ask – involved in drug metabolisation. Drugs already known to be affected include oral contraceptives and drugs used to prevent rejection of transplanted organs, but Gorski warns there may be problems whenever herbs are taken alongside any conventional drug, whether obtained by prescription or over the counter.[85]

These findings are consistent with research from the University of Naples that evaluated worldwide reports and clinical trials of interactions between the seven top-selling herbal medicines – gingko, St John's Wort, ginseng, garlic, echinacea, saw palmetto and kava – and prescribed drugs. This research showed that not only does St John's Wort lower blood concentrations of a number of drugs (including the drugs cyclosporin, amitriptyline, digoxin and warfarin) but that it can cause breakthrough bleeding and psychiatric problems when used at the same time as oral contraception. Gingko can cause high blood pressure when combined with a certain diuretic and coma when taken in conjunction with the anti-depressant drug tradozone. Ginseng can reduce the effectiveness of the blood thinning drug warfarin, prescribed in cases of deep vein thrombosis.[86] Furthermore, herbal remedies may pose a problem if taken before surgery. A review published in the *Journal of the American Medical Association* showed 'direct effects include bleeding from garlic, gingko and ginseng; cardiovascular instability from ephedra; and hypoglycemia from ginseng. Pharmacodynamic herb–drug interactions include increased metabolism of many drugs used in the perioperative period by St John's Wort'.[87]

Perhaps it would be worth taking such risks if they were outweighed by benefits, but most popular herbs and supplements appear to be ineffective. Research trials recently published in medical journals have found, amongst other things, there to be no benefit from taking saw palmetto for prostate problems, that gingko biloba doesn't improve memory in the over-sixties and that neither glucosamine nor chondroitin help arthritis.[88] [89] [90] [91] 'Each year, billions of dollars are wasted on products which are touted to achieve results that are far too good to be true,' says Carl Bartecchi, Clinical Professor of Medicine at the University of Colorado. 'The popularity of these products is fostered by the unreasonable expectation that they will somehow overcome scientific implausibility and mystically achieve the results they claim. That these herbs and most supplements don't work shouldn't be a surprise. Most never did have good scientific studies to support the claims of their effectiveness.'[92]

Even those herbs for which there has been a small amount of supporting evidence may carry risks associated with their therapeutic effects. Conventional hormone replacement therapy works by having an estrogenic effect on the body and it is thought that herbs such as black cohosh and red clover, prescribed for use in the menopause, might work because they too have estrogenic effects. It is therefore no surprise if, just as the synthetic estrogen in HRT might increase the risk of estrogen related cancers such as breast cancer, herbal remedies might also carry that risk. Once again, because black cohosh and red clover are unregulated and 'natural', it doesn't mean they are any safer than estradiol, the synthetic estrogen used in HRT. In any case, black cohosh may be of little use in treating troublesome menopausal symptoms.[93] It is now under scrutiny after several reports of liver damage associated with its use, some severe enough to require a liver transplant.[94]

Herbal medicine is the perfect example of the sceptics' favourite cry – if it worked and were safe, it wouldn't need to be alternative. It would be part of the mainstream pharmacopoeia, just as so many substances found in nature already are. In the meantime, herbal medicine should be subject to the same evidence-based regulation as are orthodox pharmaceuticals. And until it is, we shouldn't swallow it.

Chapter 5

Junk Science – with new, added junk

The electro-galvanomic machine stands in the medical gallery of the Whipple Museum of the History of Science in Cambridge. Designed in the nineteenth century to administer 'medical galvanism' in the form of a series of electric shocks, it was said to be of the 'greatest advantage in Liver and Nervous Complaints, Rheumatism, Sciatica, Tic Doloureux, Paralysis and a Variety of Diseases arising from an Imperfect Action of the Secretions'.

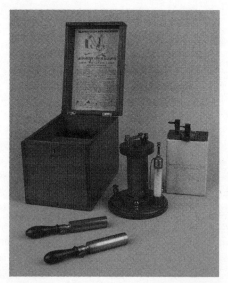

The electro-galvanomic machine at the Whipple Museum of the History of Science, Cambridge, Wh.5813

One of many similar devices produced at the time, from 'Dr S B Smith's Torpedo Magnetic Machine' to the 'Electreat', it was marketed just as electrification was transforming lives at work and at home, generating in popular culture 'electric fantasies' of infinite power, health and beauty. If we could harness this power to light our homes, surely we could also extract from it powers to provide us with health and energy? These machines offered reassurance too. 'Electrotherapy, by promising that the body could improve as a physical entity with the application of electricity, quieted fears about encroaching electric modernity', suggests Carolyn Thomas de la Peña in *The Body Electric.* [1]

She describes how in the century between 1850 and 1950 the nature of such apparatus evolved, mirroring developments in mainstream science and technology. In the twentieth century the shock-giving kits gave way to hugely popular electro magnetisers such as the I-ON-A-CO, a horse-collar style device claimed to be able to cure disease by magnetising the body's iron. Devices then emerged that reflected wider social change. Pulvermacher's Electro Galvanic Chain Belt and Electro Conducting Suspensory attachment electrified the genitals of men who feared they couldn't keep up with the sexual demands of the New Woman of the 1920s. The lady herself was plugged into the Electropoise, said to impart health-giving oxygen into the body ('You will wonder at the good results!') or the Violet Ray machine which emitted heat and a purplish light which purported to treat insomnia, headaches and a sallow complexion, as well as being a cure for tuberculosis.

Following the discovery of X-rays and radioactive materials, around the turn of the twentieth century radium-water drinks were produced with names like Vitalizer, Radithor and Vigoradium. These were said to renew energy and to be a cure for neurasthenia, the chronic fatigue syndrome of its day. Five glasses a day of what was known as 'liquid sunshine' would 'increase your income by keeping you and your family fit . . . with the alertness and vitality possessed only by those people whose bodies are functioning perfectly'. Only in the 1930s did the hazards of ingesting radioactive materials become evident. Following a number of deaths from radiation poisoning after drinking 'liquid sunshine', it was taken off the market.

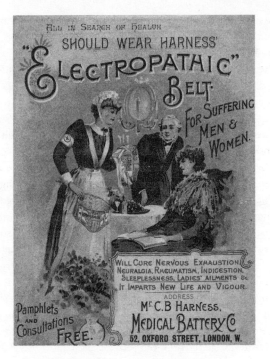

Electric fantasies of health and beauty

By the end of the Second World War the consequences of over-exposure to radioactivity had become horribly apparent and the rational fear of fallout had well and truly set in.

Given complementary and alternative medicine's professed commitment to all things natural and organic, perhaps we ought to be surprised to discover that many of these mechanical devices have travelled through time to be part of CAM today. Paradoxically, the holistic movement so hostile to the orthodox 'biomedical model', which it accuses of mechanistically isolating symptoms instead of treating the whole person with 'natural' cures, likes nothing more than a nice bit of hardcore hardware. Gadgets seemingly designed by Wallace and Gromit, with flashing lights, dials, gauges and meters, are promised to both diagnose and cure. The latest incarnations may involve lasers, or be computer-based and produce impressive readouts giving patients tangible take-home evidence of deficient Qi.

Bob McCoy calls himself 'a humble debunker of medical quackery'. In his entertaining volume *Quack! Tales of Medical Fraud from the Museum of Questionable Medical Devices* he provides this helpful checklist of how they can be identified:

- The machine apparently uses little-known energies that are undetectable by ordinary scientists
- The device can diagnose or cure people living miles away
- The machine has a convoluted yet 'scientific sounding' name
- The device was invented by a 'world famous' doctor that no one has ever heard of
- The manufacturer isn't exactly sure how or why the apparatus works
- The machine has bright lights that serve no practical purpose
- The device has knobs and dials that don't work
- The gadget shakes, rattles, rolls, sucks or warms your body
- To get results, patients must face a certain direction or use the machine only at unusual times
- The device can cure just about anything
- You're supposed to use the machine even if nothing's wrong with you
- A government agency has outlawed the device
- The machine is available only through the mail or at special outlets
- You never find the contraption at an ordinary doctor's surgery or hospital[2]

So let's fasten our magnetic seatbelts as we tour the bizarre world of junk science and some of its associated junk, from the low-tech laughable to the downright dangerous.

Energy medicine

Alternative practitioners always aspire to sound bang up to date. I have lost count of the times I've seen the phrase 'in the 21st century' affixed to presentations of purportedly two-thousand-year-old healing methods. As well as brandishing their link with the mystical wisdom of the ancients, alternative practitioners make much of their grasp of technology, or at least seek to give that impression by using plenty of techno-speak.

The notion of 'energy medicine' provides the opportunity to do all this and more. If resources are scarce, much can be achieved simply by tuning into a higher power, as does New Zealand reiki practitioner Robin Rodgers. 'Up in the sky God has a big radio station with lots of dials, all under the umbrella of energy. I can switch into the Reiki energy (by thought) and give it to you (through an attunement process) to enable you to tune into the Reiki energy. I can also tune you into the [variation of Reiki] energy because I know the special attunement for that, and now I can tune you into the new Aquarian energy by a very simple downloading attunement process.' Ms Rodgers' healing approach 'treats your body like a computer and re-programmes it . . . we are constantly amazed by the body's ability to pick out the Priority Correction (this is not always what we think it should be) and then find the emotion and the cellular memory associated with the DIS-EASE or problem.'[3]

Wind instruments

Sometimes you simply couldn't make it up. The US based 'Gentle Wind Project' distributed small plastic cards called Healing Instruments, decorated with coloured lines resembling elaborate bar codes, electrical circuits and I-Ching-style hexagrams. The healing effect is supposedly achieved by holding the cards in the prescribed way. The group claimed to distribute the cards worldwide including to people affected by physical and mental illness as well as natural disasters. The cards were not exactly sold, but 'given as *gifts* to those who make donations to TGWP'.

Their website carried the customary testimonials of success in practice: 'By having The Gentle Wind Project's Healing Instruments available for my clients' use and benefit I'm able to offer clinical interventions with a 95% effectiveness rate, unheard of in the helping professions!'[4] With names like 'Soft Sleep', 'Integrated Space' and 'Balance System' these cards have something in common with the 'clearing essences' room sprays. But in 2006 a Maine court order dissolved the organisation and banned it from making false claims about its products. Its assets are being distributed as restitution to customers.[5]

May the force be with you

How much better than gripping a decorated credit card is the chance to get hold of something with a power everyone can feel. The magnet therapy industry is estimated to have an annual global value of more than a billion dollars.[6] Magnets are sold in the form of wrist and knee bands, insoles, bracelets, mattress pads and dog bowl coasters. They are often promoted as cures for pain, especially in the neck and back, but there are claims that magnetic healing can treat everything from cancer, HIV, psychiatric disorders, stress, infections to multiple sclerosis; that they can increase energy and prolong life; stimulate the immune system and improve your golf swing. Treatment can last for minutes or months.[7]

For hundreds of years the power of magnetism had been suspected to be potentially therapeutic, but it was only with the advent of powerful carbon-steel magnets in the eighteenth century that magnet therapy really caught on. Franz Mesmer, an Austrian doctor, used magnets to treat a young mentally ill woman in 1774. Mesmer then discovered he obtained the same results without the magnets, but rather than putting this development down to the placebo effect, he decided it was his own power of healing which influenced what he imagined to be the 'universal fluid' between him and his patients.[8] He called his healing power 'animal magnetism' and moved to Paris where he set up a lucrative practice.

Mesmer claimed that health could be achieved by enabling a free flow of vital energy through the body and, like so many modern alternative practitioners, suggested that illness was caused by block-ages in that flow. His deployment of his own animal magnetism involved something similar to hypnosis. He would stare into his patients' eyes, make circular passes over their bodies with his hands and press on what is called in anatomy the hypochondriac region, an area that traverses the abdomen immediately below the ribs. There was so much demand for his treatment that Mesmer devised a group therapy system whereby he could transmit his power to many customers at a time. He would move around the room wearing brightly coloured robes and transmit his healing energy to his patients via a vessel known as a baquet, full of what he claimed was 'magnetised water'.

In Mesmer's Magnetic Institute patients would sit with their feet in a fountain of supposedly magnetised water while holding cables attached to supposedly magnetised trees. These and other activities such as hypnotism shows led to allegations of fraud, following which King Louis XVI set up a Royal Commission to investigate Mesmer's claims. Benjamin Franklin was one member of this Commission which concluded that there was no evidence for the existence of 'universal fluid' and that 'the practice of magnetisation is the art of increasing the imagination by degrees.'[9] Mesmer became the subject of ridicule and left Paris, never to be heard of again. But

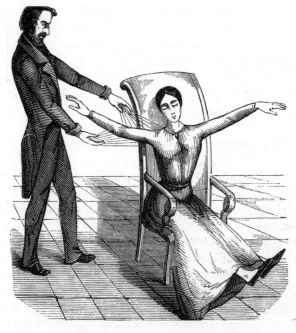

Mesmerist using animal magnetism

he lives on in the language: to be 'mesmerised' has come to mean to become transfixed and unaware of one's surroundings or to have been put in a hypnotic trance.

Today's magnet industry still pushes the same idea that magnetism can promote health by exerting some kind of positive power on the body, particularly on the blood. After all, didn't we learn something about the earth's magnetic field in physics lessons? And of course we all know that blood contains iron. So it sounds feasible when we read, in an advert for Green Foam Magnetic Insoles (£12.95 a pair) that 'magnetic therapy is the application of magnetic fields on parts of the body to speed healing, relative pain and inflammation, and improve bodily functions. The clinical benefits of magnetic therapy *being researched* [my italics] include: pain reduction abilities; healing capabilities to bone, tissue, muscle and nerves; chronic disease prevention and reversal; antibiotics to which microorganisms cannot build a tolerance; effects of more available oxygen and thus more energy'. Note the use of the words 'being researched' as a way to support unsubstantiated claims. Red flag or what?[10]

Visions of still redder flags accompany the 'LadyCare', a 'small, powerful, discreet device that attaches firmly to your underwear over your abdomen' to treat menstrual disorders. According to its manufacturers Magnopulse, 'medical researchers believe that with LadyCare more oxygen-rich blood reaches the muscles of the uterus, helping them to work more effectively. This should stop the build up of escess [sic] acids and reduce cramps.' Sorry to disappoint, sisters – but those unnamed, unpublished medical researchers are wrong. The iron in blood is contained in haemoglobin molecules and is not *ferromagnetic* like the metal of your fridge door but very slightly *diamagnetic*, which means it is repelled, not attracted, by magnets.

Magnopulse is one of the many manufacturers who capitalise on this widespread and false belief that the iron in blood is ferromagnetic. If magnets had any real effect on blood then no human being would survive being scanned by a Magnetic Resonance Imaging machine, which contains magnets many times the strength of those in Green Foam Insoles. As Christopher Wanjek points out in his book *Bad Medicine*, if magnets affected blood flow they would cause the delicate veins in your body to explode when you are in an MRI scanner.[11] The heavy-duty magnets in MRI scanners are strong enough to have been known accidentally to suck in hospital equipment including wheelchairs and floor polishers. The most tragic recorded case was that of a six-year-old American boy who died during an MRI scan in 2001. He was sedated in an MRI machine after a brain operation, when his oxygen supply failed. The anaesthetist ran outside for an oxygen tank but failed to notice it was made of steel. As he returned, the tank shot out of his hands, hitting the boy in the head and killing him instantly.[12]

In any case, the therapeutic magnets on today's market are generally so weak that their magnetic field is hardly able to penetrate their wrappings, as physics professor Robert Park discovered when he bought a 'Thera-P' magnetic armband for $39.95. To assess how far the field extended he took one of the magnets from its Velcro pouch and stuck it on a steel filing cabinet. He placed sheets of paper between the magnet and cabinet until the magnet could no longer support itself and fell off. The field extended through no more than ten sheets of paper, a measurement of just one millimetre: 'the field of these magnets would hardly extend through the skin, much less penetrate into muscles,' says Park. When he put the magnet back in the pouch the strength of the field was so reduced that it could not pick up a paper clip. 'Even a pair of scissors on my desk had enough residual magnetism to do that.'[13]

Beyond doubts about the limits of magnetic power, there is no good evidence that magnets work, despite the marketing hype. Research in this area is notoriously unreliable, as researchers Leonard Finegold and Bruce L Flamm remark in a review of magnet therapy research published in the *British Medical Journal* in 2006. They say the reason many controlled experiments are suspect is because it is so difficult to blind subjects to the presence of a magnet. Patients given real magnets can easily detect them when the magnets stick to keys in pockets, or when they sense the magnetic drag when near ferromagnetic surfaces. When research is actually carried out in a way that prevents people knowing whether or not they are being treated by a magnet, there is no evidence they are any more effective than a placebo. Finegold and Flamm conclude that 'patients should be advised that magnet therapy has no proven benefits. If they insist on using a magnetic device they could be advised to buy the cheapest – this will at least alleviate the pain in their wallet'.[14]

What therapeutic magnets certainly do attract is a high price tag. The coin-sized LadyCare device may have no more power than a fridge magnet, yet it costs £19. A Magnopulse magnetic 'pet coaster' for your dog is £20. You place it in the pet's water bowl because, they say, 'magnetic treated water is more natural' and 'using a pet coaster ensures that your pet receives maximum benefit from their drinking water . . . given the choice your dog will always choose magnetic water'. All complete nonsense because, funnily enough, there is no such thing as magnetised water. Water, like blood, is very slightly repelled by a magnetic field.[15] But Magnopulse aren't bothered: they're busy promoting the 'SleepCare' magnetic underlay, which they say raises the body temperature so as to encourage 'a better and more restful night's sleep, helping with backache, neurological disorders, stress and strokes' and retails for a strongly repellent £229.

Magnopulse magnets are on a roll in the UK since they became available on the National Health Service in the UK in 2006. *The Sunday Times* reported that NHS accountants were so impressed by the cost-effectiveness of Magnopulse's magnetic leg wrap, 4UlcerCare, they decided to put it on the official list of items that can be prescribed by NHS doctors. This means pensioners and others exempt from prescription charges can get them for free. The Prescription Pricing Authority said that the magnets, purchased from Magnopulse at the knock down price of £13.80 per wrap instead of the list price of £29, 'would save money on bandages and nurses' time by healing the wounds.'[16]

But, as Finegold and Flamm demonstrated, there is just no support for the idea that these magnets have any positive effect on healing. No published research exists on the 4UlcerCare wrap beyond a customer survey of a small number of existing Magnopulse customers by Dr Nyjon Eccles, a 'bioenergetic' general practitioner with a Harley Street practice.[17] Yet this was apparently good enough for the PPA. Such state sponsored support of magnet therapy has angered campaigning sceptic and University College London's Professor of Pharmacology David Colquhoun because 'it seems to suggest that whatever data are submitted by the manufacturer are taken at face value, without any attempt to evaluate their quality. I dare say [the PPA] could save even more money by removing all effective treatments.'[18] Since the PPA approval of Magnopulse's products the Office of Fair Trading investigated Magnopulse and found that a number of the company's advertising claims were misleading. The company has now stopped claiming, amongst other things, that their products have a therapeutic effect and that magnetic products magnetise or ionise water.[19]

Short circuit

The electro-quacks of the nineteenth century might have been able to get away with it, but thankfully it has become more risky these days to make expansive and unsubstantiated claims about a machine's ability to treat or cure an illness. This must be one of the reasons why many devices are now found in the area of 'electrodiagnostics', with machines claimed to be able to detect energy imbalances, and other nebulous disorders. These new devices are also the perfect way to provide names and explanations for the types of conditions sufferers fear might be dismissed by conventional practitioners as being 'all in the mind'.

One of the most ubiquitous devices – called variously the Vegatest, Avatar, Interro, BioMeridian, Omega Acubase and the Meridian Stress Assessment System – is the EAV or 'electrodermal screening' machine, EAV standing for 'Electro Acupuncture according to Voll'. The German acupuncturist and physician Reinhold Voll invented the process in the 1950s, but today EAV is employed by the full gamut of alternative therapists, including chiropractors, homeopaths and alternative nutritionists. The patient is wired up to a galvanometer, which measures the electrical resistance of the skin, and a low voltage circuit is created. A pen-sized probe – one version is called the 'Epic Stylus' – is pressed on the skin at various acupuncture points, and any variations in the current are registered on a gauge from nought to one hundred.[20] Readings over fifty-five are said to suggest inflammation of the organ associated with the acupuncture point being tested, readings below forty-five are supposedly a sign of organ stagnation or degeneration.[21] Some machines feature sounds and flashing lights and the latest models are linked to a computer screen so you can see your diagnostic readings in living colour.

Electrodermal screening is also used for diagnosis of food or environmental allergies and intolerances. 'Dr Voll found that the electrical characteristics of an acupuncture point varied as a substance a person was sensitive to, depleted in or would benefit them was placed near them. This basically gave him a way of asking the body questions such as "do you have a sensitivity to bananas?" by placing a banana near the person and measuring the reaction of an acupuncture point', according to the Hampshire based Biotech Health and Nutrition Clinic. The Biotech Clinic offers a comprehensive screening

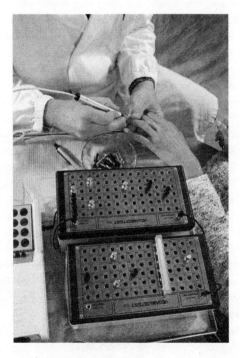

Vega testing in action

session giving a 'full picture of energy balance, organ functioning including the toxic triangle test' for a mere £190.[22]

In addition, the patient can hold a remedy or supplement to see whether introducing it into the electrical circuit elicits a positive response from the machine, which in turn will elicit a positive sales opportunity for the alternative practitioner, very welcome given the considerable outlay required to purchase an EAV machine. Salespeople are coy about the cost, preferring to print 'price on application' on their advertising material, but I found one supplier who admitted to selling one for £13,300.[23]

There has been little independent research into such devices. The BBC's *Inside Out* programme investigated VEGA testing in 2003 and their reporter Chris Packham took the test in three different Holland & Barrett outlets. His 'allergy' results totalled over thirty-three different foods, including staples like wheat, potatoes, milk, tomatoes, tea and coffee. Of these thirty-three items, the testers agreed on only two – cheese and chocolate. The reporter 'was also advised by Holland and Barrett staff to take a total of twenty different vitamins and minerals. But again, the testers can't seem to agree and all three advised different supplements.'[24]

Examination of a Vegatest kit commissioned by Quackwatch suggested that the EAV score is, knowingly or not, determined by the practitioner. The figure registered simply depends on how hard the probe is pressed on the patient's skin, leading the experimenter to conclude that the machine was 'simply a resistance-measuring instrument'.[25] In other words it is the therapist, not the EAV machine, who determines your organ function, your need for any particular alternative remedy or sensitivity to bananas. Dr Stephen Barrett of Quackwatch believes that 'EAV devices should be confiscated and that practitioners who use them should be delicensed because they are either delusional, dishonest, or both.'[26] He lists attempts by government agencies around the world to stop EAV and related practitioners from making false claims and misleading patients both in advertising and in practice, including the following examples.

In the UK the Advertising Standards Authority has ordered the withdrawal of adverts for Vegatests that make misleading claims such as a 'complete test for hidden problems' or that claim a 'Bioresonance

Therapy' device would help with headaches, overweight, tiredness, IBS and arthritis. In Australia there have been moves to stop claims that the 'EQ4 Computerized Electrodermal Screening Device' could both detect allergies and create homeopathic remedies, and a naturopath who used a MORA device was convicted of manslaughter in connection with the death of an infant he claimed to have cured of a heart defect. And in the US a chiropractor was disciplined for 'fraud and misrepresentation' after he falsely told a haemophiliac man with HIV/AIDS that the 'Interro' EAV device had shown him to be clear of the virus. The patient subsequently died of AIDS complications, but not before the virus had been passed to his wife and newborn daughter, the couple having gone ahead with a pregnancy because the chiropractor's advice led them to think it was safe to do so.[27]

Testing testing

The language of alternative medicine may suggest a rejection of the eighteenth-century Cartesian model of the body as a machine, yet CAM has no qualms about describing the body as a computer. 'Energetic medicine is the oldest, and possible the most effective, form of healing', claims the UK firm QQS, which markets the Quantum SCIO – the Scientific Consciousness Interface Operations System – also known as the QXCI, the Quantum Xxroid Consciousness Interface. 'Never before has there been so much attack on our "Template Software" creating disruption to our natural processes. The QXCI can be considered as Naturopathic Technology'.[28]

This latest addition to the armoury of bioenergetic merchandise, the Quantum SCIO or QXCI was invented by an unemployed maths teacher and cabaret artiste. US-born William Nelson was indicted on nine counts of felony fraud in 1996 since when he fled to Hungary.[29] It is claimed to be able to *treat* as well as diagnose a multitude of disorders, allergies and imbalances. The client is linked to the machine via electrodes strapped to the head, wrist and ankle, whereupon 'tri-vector energy is compared with 8000 items stored on computer. This process takes three minutes while the client sits quietly, they are unlikely to feel anything. An automated statistical analysis is then performed and the results are presented as a coloured table'. QQS points out that *scio* is Latin for 'I know', which is cute.

The treatment phase is where things really start to take off. Built-in therapies include allergy desensitisation, craniosacral therapies, electro acupuncture, homeopathy and even brain wave adjustment; all carried out by the computer, not the practitioner. Instead they are 'automatically performed by passing electrical information [low level radio frequencies] through the head, wrist and ankle connections' and most 'take a matter of seconds to perform but can be set to run in continual background mode or [my itals] *to continue to be performed even when the client has left* (physical connection is not necessary)'.[30]

Even in notionally sympathetic energy medicine circles there has been a buzz of dissent about the QXCI. 'It works simply by entering data into a screen, which is fed into a weighted random number generator which in turn displays results,' writes one critic. 'When it comes round to treating, buttons are just clicked on a screen with no programming beyond the graphical displays of saying what it is treating.' Remarkably, given these reservations, things may not be all bad: 'we are not saying that the QXCI does not have a place or a use or is ineffective. Repeated use by many practitioners bears testament to the fact that some useful interpretation of results can be gained, especially by not entering data (*sic*).'[31]

You are probably beginning to have a few doubts about the SCIO/QXCI. But an alternative practitioner might find other, more persuasive reasons for acquiring one. For a start, as another review of the machine says, it will enable them to 'work with a multitude of therapies without having to know much about them or have to touch the client'.[32] It may be expensive (£10,700 in the UK and $18,000 in the US) but bear in mind that 'technology attracts clients and charges are high for practitioners who use state of the art assessment and therapy systems. Typically practitioners will charge an extra £20 per session hour above normal rates,' according to the UK QXCI sales company QQS.[33] Another UK supplier, NutriVital Health, promises even better returns, suggesting that 'fee rates for energetic screening are on average 75% higher than for unassisted consultations in the fields of nutrition/homeopathy/naturopathy'. NutriVital kindly provides a financial illustration. 'Fees vary geographically but an average hourly rate of £40 for complementary health consultations might reasonably be raised to £70.

Assuming 15 hours per week of consultations this represents: **INCREASE** in income of: £450 per week or £1,950 per month or £23,000 per year' (NutriVital's bold caps).[34] Considering all those therapies that can be performed 'even when the client has left' this must add up to an alluring package for the hitherto 'unassisted' CAM practitioner.

Long distance information

However tempting the chance of extra income from performing their own diagnostics, many therapists prefer to send samples away to commercial altmed laboratories for testing, many of which do controversial and unconventional tests. Their results seem to generate suspiciously frequent diagnoses of controversial conditions like environmental allergies, food intolerances, candida or Lyme disease. As we have seen in other complementary medicine contexts, they may employ confusing technical language, as does one US laboratory, Genova Diagnostics, used by alternative practitioners from all over the world. This company offers what it calls 'innovative' testing which 'assesses the dynamic inter-relationship of physiological systems, thereby creating a more complete picture of one's health, unlike traditional allopathic testing, which is more concerned about the pathology of disease'.

Once again, Quackwatch provides a useful list of both dubious diagnostic tests and questionable laboratories, but this is not exhaustive and covers only those based in the US.[35] It does not, for example, list Bionetics, a company based in Camberley, Surrey, offering mail order hair analysis. You are invited to send them a few strands of your hair which are then analysed by instruments they claim 'read energy signals and identify the individual stresses and nutritional requirements'. The firm says that by 'reading your body's energy field' in this way they 'can identify those specific underlying factors which are impacting on your personal health'. Bionetics then creates 'the action plan that you need to adopt to eliminate the inhibiting factors, rebuild your natural defences and speed up the healing process'.[36]

The test, together with an online report, costs £48. The prescribed 'action plan' consisting of a personalised list of recommended herbs, supplements and homeopathic remedies, all of which can be purchased direct from Bionetics, may cost considerably more. In a sample report for prospective customers, we find a diagnosis of 'fungus', as well as of a variety of food intolerances – hazelnut, aubergines, chocolate, tuna and rice. Besides telling the patient to avoid these foods the report recommends a ninety-day 'action plan', which requires the purchase of eight different products from Bionetics at a total cost of nearly £150. One of the recommended remedies, the South American herb pau d'arco, is described in the Natural Medicines Comprehensive Database to be of unknown effectiveness in treating infections (including candida, which one supposes is what Bionetics mean by 'fungus') and to be potentially unsafe even in small doses. There is, however, evidence that pau d'arco is unsafe in high doses, when it can cause severe nausea and vomiting, dizziness and anaemia.[37]

Bionetics offer no evidence for their claims beyond testimonials. These feature Bart of Berkshire, who says 'since I found Bionetics I haven't looked back – the first month was hard as it involved a whole dietary life change, but now I have energy and enthusiasm (my spark has returned), my weight has stabilised'. Claire of Aberdeen writes 'I am feeling so much better and cannot believe how much my immune system has improved. That's almost three months without a virus, which is a huge achievement. My energy levels have also increased which is great. I actually feel as if I have slept when I get up in the morning, which is a revelation!'

Such testimonials, however, cut no ice with the Advertising Standards Authority, which polices the advertising codes in the UK. In response to a complaint from the public, the ASA decided in 2006 that Bionetics' claims that their test could 'establish whether or not your body has become intolerant to 123 of the most common problem foods and ingredients' could not be justified and advertising containing this claim was banned.[38]

The con is on

In recent years the foot spa, which used to be the first Christmas present to be shoved to the back of the wardrobe come January, has come out of the closet as a 'detox footbath'. The American naturopath Dr Royal Rife invented what he claimed was an electromagnetic detoxification device in the 1930s and it is this 'Rife machine' that was introduced to Europe in 2001 in the form of a foot spa.[39] Under names like Ion Cleanse, Aqua Chi or Bio Energiser, it retails for as much as £1,500.

The Nightingale Complementary Health and Beauty Clinic in Earls Barton, Northamptonshire, charges £35 for a half-hour session with a Hydra Detox Machine and can't praise it too highly. Their claims deserve to be reproduced in full. They say that people who use the Hydra Detox Machine have 'reported alleviation of: Migraines, Hypertension, Iodema [sic], Depression, Fatigue, Psoriasis, Eczema, Arthritic pain, Asthma, IBS, PMS, ME, MS, Diabetes and Acne'. They also say 'clients have also reported improvements in Sleep patterns, Skin texture, General metabolism, Liver & kidney function, Alertness, Hair condition, Blood sugar, Acidity levels, Blood pressure, Cholesterol levels, Thyroid imbalance'.[40]

The Hydra Detox foot spa contains an electrical element – an array – that reacts with water and added salt to create an electrolysed solution, which the manufacturers say creates an electromagnetic field that 'stimulates your body bio-energetically via the feet or hands'.[41] The idea is that 'the footbath moves toxins through the body, causing the water in the footbath to discolour, and improves the body's own lymphatic drainage system'.[42] The spa is switched on and the feet are immersed. The water in the footbath turns a brackish colour, which users and practitioners believe shows the body's toxins are being released through the pores of the feet. The darker the water the better. The shade may even be interpreted as part of more detailed diagnostics, with a yellow or greenish tinge signifying detoxification from the kidney, bladder and urinary tract, dark green showing toxins from the gallbladder and white foam representing toxins from the lymphatic system.[43] As one user attests 'you wouldn't believe the gunk that this twenty minute footbath draws out of the body, which seems to confirm that detox is our real issue'.[44]

In fact the water would change colour whether or not your feet or hands were in the footbath. That nasty brown stuff is nothing more than rust, produced by the combination of electrifying array, water and salt. It's the same trick as perpetrated by those 'Hopi' ear candlers and yet another example of *post hoc ergo propter hoc* (after this therefore because of this), one of the core tricks used in alternative medicine. Whilst the small print may mention in passing that 'some of the colour change does occur due to the fact that salt is added to the water' it is still this effect which is consistently headlined and believed by users to be a sign that the foot bath is detoxifying. And if that weren't enough, if a session in the detox footbath irritates the skin or produces a rash, so much the better, as 'it is caused by toxins being removed from the body'.[45]

Shut down

Whilst some of these devices may seem ludicrous and aimed at the gullible, reliance on junk science by those in extremis can be dangerous. In 2003, Reginald Gill, a self-styled 'wellness practitioner' who ran an alternative medicine clinic from his bungalow in Poole, Dorset, was jailed for twelve months for falsely claiming a machine he sold a dying man would 'kill' his cancer.

Stephen Hall was diagnosed with advanced pancreatic cancer in 2002. He was forty-three. The cancer was inoperable but he was offered chemotherapy that his doctor hoped could extend his life by as much as a year. Devastated by the news that he had a terminal illness, Hall turned to Reginald Gill and the IFAS High Frequency Therapy machine Gill had bought from an Australian supplier for £200. Gill gave Stephen Hall a leaflet advising cancer was a metabolic disorder that could be treated and reversed. He discouraged the use of chemotherapy, saying 'if you have chemo, you'll go home in a box'. He said he'd used the IFAS machine to treat everything from cancer to baldness and described it as 'magic'.[46]

Gill also told Hall to stop taking the morphine painkillers he had been prescribed and to follow a highly restricted diet. He was instructed to drink a mixture of various supplements, pineapple juice and organic hemp seed oil before treatment with the IFAS machine, which Gill claimed would enable the body to use the nutrients in the drink to heal the cancer. After the first two-hour session with the IFAS Gill said 'I've got it. I've killed the bad cells. It's just the pancreas that needs more work'. Stephen Hall arranged for four more £75 sessions with the IFAS machine, which Reginald Gill subsequently sold to him for £2,500 – £2,300 more than he originally paid for it.[47]

Tragically, just ten weeks after diagnosis, Stephen Hall died of his illness. His family was brave enough to make an official complaint that led to the trial and conviction of Gill following two charges under the Trade Descriptions Act. Sentencing Gill the judge told him 'when I first began to hear this case, I considered and wondered if you might be a sincere but misguided eccentric. As the evidence has developed I did change my view. I find that in your evidence you demonstrated a disturbing level of callousness and cynicism towards Mr Hall.'[48]

The IFAS machine produces a low voltage electric current together with warming violet rays. I have seen no claims that the machine cures cancer, but its manufacturer does describe it as 'the complete answer to most health problems'.[49] Australian 'electropractitioner' David Hannaford (who attaches the qualification Adv. Dip. IFAS to his name) says the electrical forces administered by the IFAS machine 'stimulate and strengthen vital organs, develop the body, steady the nerves, and activate sluggish or painful muscles, purifying and causing the flow of warm rich blood through the treated part'.[50]

The IFAS machine and its accompanying claims are virtually identical to those made in the nineteenth century for the electro-galvanomic machine, now properly consigned to the medical gallery of the Whipple Museum of the History of Science in Cambridge. We are back where we started. In the world of alternative medicine, history is junk.

Chapter 6

Bad Backs: Are you being manipulated?

The 'bad back' is the scourge of our age. It costs the UK an annual £6 billion in benefits, treatments and lost production – nearly 2 per cent of our gross national product. Every year half the population will suffer from back pain lasting at least a day. Eighty per cent of us suffer from back trouble during a lifetime. It is the leading cause of long-term disability, with one in eight of the unemployed saying they cannot work because of it.[1]

Anyone who has had back pain will know the debilitating impact it can have on one's entire life and how frustratingly difficult it can be to get any respite. Years ago, GPs would advise bed rest. Today they might suggest painkillers together with a little gentle exercise, but such advice seems worse than useless when your back is in spasm. And if your pain is such that you are unable even to sit at a desk, being informed by your GP that most back pain gets better of its own accord is not a comfort, but a counsel of despair.

The option of a consultation with an osteopath or a chiropractor may be more than welcome for both doctors and sufferers – both parties get to feel they are trying to make a positive difference. It is therefore no surprise to find that over three-quarters of patients who use complementary and alternative medicine have a musculoskeletal problem as their main complaint.[2] The majority of these will choose to consult a practitioner of manipulative therapies, no doubt the chief reason why osteopathy and chiropractic enjoy a level of integration into mainstream medicine surely the envy of the rest of alternative medicine. Osteopathy and chiropractic have both managed to gain huge respectability simply by being officially regulated, despite there being no official standards of efficacy for either practice. State regulation is equated with state approval, hence they go largely unquestioned and uncriticised. Furthermore, customer satisfaction is reputedly high, with many patients in no doubt that their back problems have been alleviated by these therapies.

Practitioners in both disciplines predominantly use their hands to work with bones, muscles and connective tissue. One frequently employed manipulation is the high velocity thrust, a short and sharp motion usually applied to the spine, intended to release structures with a restricted range of movement.[3] Such thrusts often produce the sound of a crack or pop, caused by the release of the accumulated gas in the fluid around the joints. It's what you hear when you crack your knuckles. This crack has no known benefit but comes as an audible sign that the patient is getting their money's worth. Chiropractors are more likely to manipulate the spine directly; osteopaths may use the limbs as levers and in this way try to mobilise the spine.

Both osteopathy and chiropractic originated in America. They were a development of the medieval folk medicine practice of 'bone setting', which employed vigorous massage to treat strains, sprains and fractures.[4] In America today there are more than forty-nine thousand Doctors of Osteopathy (known as DOs) who are trained in orthodox scientific medicine with additional training in manipulative therapies. They have the same entitlements to prescribe and perform surgery as mainstream medical practitioners and make up 20 per cent of all general practitioners in the US.[5]

The UK's five thousand or so osteopaths, however, require no scientific medical training and are therefore much more firmly established in camp 'alternative' than their US colleagues. Very few are medical doctors and many combine osteopathy with dubious practices such as naturopathy and cranial osteopathy. Since 1993 and the passing of the UK Osteopathy Act in 1992, anyone who calls themselves an osteopath must be registered with the General Osteopathic Council. In contrast, chiropractors in both the US and the rest of the world have no scientific medical training but often use the title 'doctor' and may affix DC (Doctor of Chiropractic) to their names. In the UK more than two thousand practising chiropractors are now compulsorily registered with the General Chiropractic Council. At the last count Australia had around two thousand state-regulated chiropractors for a population around a third of Britain's. In the US licensing rules for that country's fifty-three thousand chiropractors vary from state to state.

Osteopathy and chiropractic may be popular with patients the world over, but make enquiries beyond reports of customer satisfaction and there is minimal evidence to show that spinal manipulation is any better at alleviating back pain than gentle exercise. In a review of current research that enraged osteopaths and chiropractors, leading sceptic Professor Edzard Ernst concluded that there was no evidence to suggest that spinal manipulation was an effective intervention for any condition. This finding applies to both osteopathy and chiropractic.[6] Similar conclusions were reached by Dr Scott Kinkade of the University of Texas in another research review. He concluded that manipulative therapy might provide short-term benefits compared with sham therapy, but not when it was compared with conventional treatments including painkillers, anti-inflammatories, heat treatments and advice to stay active.[7] What's more, the questionable benefits may be outweighed by very real and known risks. Chiropractic in particular involves neck adjustments that can be highly dangerous. These adjustments can cause the tearing of the arteries in the neck and have been known to result in stroke, paralysis and death.

Osteopathy and chiropractic were invented, or 'discovered' as their adherents prefer to say, by a pair of determined and charismatic Americans in the late nineteenth century. Andrew Taylor Still and Daniel David Palmer didn't know each other, but they had much in common. Both could be described as chancers and fantasists who had tried and failed to make their fortunes in a variety of jobs and get-rich-quick schemes. Both saw themselves as visionaries and spiritual leaders. And in common with so many alternative practitioners, both believed they had discovered a single cause and a single cure for all diseases.

Osteopathy – the 'whole truth'

Andrew Taylor Still was born in 1828. As a young man he worked as a farmer but was always fascinated by technology and had ambitions to be an inventor. In the mid-1850s he built and opened a steam-powered sawmill and later developed a wheat-mowing machine, an improved butter churn and an anti-pollution device for coal furnaces. He supported the anti-slavery movement and is recorded as working as a hospital steward during the American Civil War, as well as serving as a combatant for the Union Army. In later years Andrew Taylor Still claimed he had been a battlefield surgeon, but no record of this exists. He returned home from duty in 1864 to suffer the terrible loss of three of his children from meningitis. Three other children had previously died of infectious diseases soon after birth. Following these appalling events he described himself as being 'torn and lacerated with grief' and thoroughly disillusioned by what he saw as the ineffectiveness and barbarity of orthodox medicine. Phrenology and mesmerism had interested him for some time, but the deaths of his children precipitated his exploration of spiritualism and his decision to become a magnetic healer.[8]

Andrew Taylor Still, inventor of osteopathy

Famously, at ten in the morning on June 22, 1874, Andrew Taylor Still had a 'prophetic vision' where he was shot, as he described in his autobiography, 'not in the heart, but in the dome of reason' where 'like a burst of sunshine the whole truth dawned upon my mind'. This truth was that 'there is no such disease as fever, flux, diphtheria, typhus, typhoid, lung-fever, or any other fever classed under the common head of fever or rheumatism, sciatica, gout, colic, liver disease, nettle-rash, or croup, on to the end of the list, they do not exist as diseases'. Instead 'all diseases are mere effects, the cause being a partial or complete failure of the nerves', brought about by misalignments of the vertebrae. If the bones could be manipulated back into alignment then the nerves would then 'properly conduct the fluids of life' and so-called 'dis-eases' or 'effects' would trouble the patient no longer. The body would be free to cure itself once the physician had removed the impediments to healing.[9]

For Still, this approach was to concentrate all his interests and passions in a single philosophy, a kind of spiritual mechanics. He never doubted he was right, since 'day by day the evidences grow stronger and stronger that this philosophy is correct'. But Still's approach displeased the minister of his Methodist church, who accused him of trying to emulate Jesus Christ in the laying on of hands. Andrew Taylor Still was branded an agent of the devil and a practitioner of voodoo medicine. By the end of 1874 he had been formally expelled from the congregation, deserted by his patients and socially ostracised. He was forced to find work as an itinerant 'Lightning Bone Setter', as he described himself, travelling from town to town to find patients.

By the mid 1880s he had coined the word 'osteopathy' – from the Latin *osteo* (bone) and *pathy* (suffering or disease) and business was booming. Andrew Taylor Still was doing so well that he was able to establish a permanent practice and, soon after, the first American School of Osteopathy in Kirksville, Missouri, which flourishes today in addition to another eighteen schools of osteopathy in the USA and seven in the UK. The curriculum at Kirkville is founded on eight osteopathic principles developed by Andrew Taylor Still, catch-all principles so vague as to be virtually incontestable, such as 'the body is a unit' and 'the body has the inherent capacity to defend and repair itself'.[10]

By the time Andrew Taylor Still died aged 89 in 1917, osteopathy was popular throughout the US and had been licensed in several states. He spent his latter years practising spiritualism and trying to perfect a system whereby meditation could be used to answer philosophical questions. A hawker of homespun homilies such as 'to find health should be the object of the doctor. Anyone can find disease', Andrew Taylor Still is venerated by modern osteopaths nearly a century after his death. He is admired by one biographer as 'a farmer, expert hunter, medical doctor, inventor, machinist, state legislator, soldier, patriot, Civil War veteran, abolitionist, feminist, temperance supporter, Freemason, father, husband, and dedicated family man'.[11]

In the years since Still's death osteopathy may have attained the highest status of all fringe medical practices, but the most recent review of the evidence for spinal manipulative therapy as a treatment for lower-back pain found 'it was no more or less effective than medication for pain, physical therapy, exercises, back school or the care given by a general practitioner'.[12]

Chiropractic's cracking tale

You may recall the lone genius I warned about in chapter two, who develops a whole new medical discipline all on his ownsome. Now meet a red flag in human form, the inventor of chiropractic, Daniel David Palmer. Born in Canada in 1845 and moving with his family to the US at the age of twenty, D D Palmer as he was known, had a variety of jobs including beekeeper, grocer and fishmonger but all the while, in common with osteopathy's Andrew Taylor Still, was developing a keen interest in magnetic healing and spiritualism.

By the 1890s Palmer had established a magnetic healing practice in Davenport, Iowa, and was styling himself 'doctor'. Not everyone was convinced, as a piece about him in an 1894 edition of the local paper, the *Davenport Leader*, shows. 'A crank on magnetism has a crazy notion that he can cure the sick and crippled by his magnetic hands. His victims are the weak-minded, ignorant and

Daniel David Palmer, inventor of chiropractic

superstitious, those foolish people who have been sick for years and have become tired of the regular physician and want health by the short-cut method . . . he has certainly profited by the ignorance of his victims, for his business has increased so that he now uses forty-two rooms which are finely furnished, heated by steam and lighted by forty electric lights . . . he exerts a wonderful magnetic power over his patients, making many of them believe are well. His increase in business shows what can be done in Davenport even by a quack.'[13] Palmer's clinic also housed what was described as the finest collection of mounted animal heads in the west, as well as a glass enclosure containing four live alligators.[14]

Just like Andrew Taylor Still, Palmer decided there must be a single cause for all diseases and was determined to find it. 'I desired to know why one person was ailing and his associate, eating at the same table, working in the same shop, at the same bench, was not?'[15] In a striking echo of Still's osteopathic revelation, the idea for chiropractic came to him 'from the other world' during a seance, when he believed himself to be in contact with a deceased physician, a Dr Jim Atkinson. Soon afterwards, inspired by what had been communicated to him by Dr Atkinson's spirit, Palmer famously attempted chiropractic's first 'adjustment'. The date, September 18, 1885, is carved on Palmer's memorial stone at Port Perry, Ontario, his Canadian birthplace.

Spinal tap

He later gave this account of what happened: 'Harvey Lillard, a janitor, in the Ryan Block where I had my office, had been so deaf for 17 years that he could not hear the racket of a wagon on the street or the ticking of a watch. I made enquiry as to the cause of his deafness and was informed that when he was exerting himself in a cramped stooping position, he felt something give way in his back and immediately became deaf. An examination showed a vertebra racked from its normal position. I reasoned that if that vertebra was replaced, the man's hearing should be restored. With this object in view, a half-hour's talk persuaded Mr Lillard to allow me to replace it. I racked it into position by using the spinous process as a lever and soon the man could hear as before. There was nothing "accidental" about this, as it was accomplished with an object in view, and the result expected was obtained. There was nothing "crude" about this adjustment; it was specific, so much so that no Chiropractor has equalled it.'[16] Daniel David Palmer decided to call his janitor's supposed spinal irregularity a 'subluxation', a term borrowed from orthodox medicine, where the word refers to a partial dislocation of a joint, readily visible on X-ray. The correction of chiropractic subluxations was to become central to chiropractic and remains its most contentious practice.

As with so many of the stories of chiropractic's early days, Lillard's daughter disputed D D Palmer's account. She said her father told her that he was telling jokes to a friend in the hall outside Palmer's office and Palmer joined them. When Lillard reached the punch line Palmer laughingly slapped Lillard on the back with the hand holding the heavy book he'd been reading. A few days later, Lillard told Palmer that his hearing seemed better. Palmer then decided to explore manipulation as an expansion of his magnetic healing practice. Lillard reported 'the compact was that if they can make [something of] it, then they both would share. But it didn't happen'.[17]

Instead, a few weeks later, Palmer claimed, he gave 'immediate relief' to a woman with chronic heart trouble by performing a vertebral adjustment. 'Then I began to reason if two diseases, so dissimilar as deafness and heart trouble, came from impingement, a pressure on nerves, were not other diseases due to a similar cause? Thus the science (knowledge) and art (adjusting) of Chiropractic were formed at the same time. I then began a systematic investigation for the cause for all diseases and have been amply rewarded.'[18] Palmer's idea was that pressure on the nerves would restrict psychic energy, just as pinching a garden hose slows the flow of water to a flowerbed; if you remove the pressure the flow is restored to normal and the garden blooms.

Palmer named his method chiropractic from the Greek *cheir* (hand) and *praxis* (action). He asserted that manual adjustments of the spine enabled the free flow of what he now called the body's 'Innate Intelligence', meaning the life force or vitalistic energy beloved of so many alternative methods today. 'Functions performed in a normal manner and amount result in health,' he said. 'Diseases are conditions resulting from either an excess or deficiency of func-tionating [*sic*]. The dualistic system – spirit and body – united by intellectual life – the soul – is the basis of this science of biology, and nerve tension is the basis of functional activity in health and disease . . . chiropractors correct abnormalities of the intellect as well as those of the body.'[19]

The use of such language, with its expression of quasi-religious ideas concerning spirit and soul, was intentional. Moreover D D Palmer was emphatic that chiropractic was a religion and that he was its leader. In a letter to an associate he wrote 'I occupy in chiropractic a similar position as did Mrs [Mary Baker] Eddy in Christian Science. Mrs Eddy claimed to receive her ideas from the other world and so do I. She founded a religion thereon, so may

I. I am THE ONLY ONE IN CHIROPRACTIC WHO CAN DO SO . . . we must have a religious head, one who is the founder, as did Christ, Mohammed, Joseph Smith, Martin Luther and others who have founded religions. I am the fountainhead. I am the founder of chiropractic in its science, in its art, in its philosophy and in its religious phase.'[20]

At first D D Palmer wanted to keep his invention of chiropractic a secret. His son, Bartlett Joshua Palmer (known as B J and born in 1882) wrote that it was D D's policy 'to let no one see him give an adjustment; no mother could see the child take one; no husband see his wife; no patient see another. It is a fact that one day Father saw his patient "peeking" in a mirror to see how it was done. The mirror was taken down at once and women ever after dressed in mirrorless rooms'. B J claimed that from an early age he recognised the potential benefits of popularising the method but that his father disagreed, indeed 'the earliest words that I recall passing between Father and myself were on the contention that it was wrong to bottle up Chiropractic to the few. If it was what he claimed it, the world needed it; needing it they must get it'.[21]

It was not long before D D felt able to open the Palmer School of Cure in Davenport, Iowa in 1897. According to the historian Ralph Smith, the chiropractic course ran for three months; the student learned how to adjust spines and received a crash course in medicine by reading 'Dr Pierce's Common Sense Medical Adviser in Plain English, or Medicine Simplified'.[22] The Palmer School 'began with a tuition fee of $500, dropped it to $300, then raised it to $450 with a $50 discount for cash. The only admission requirement was the ability to pay the fee'. By 1905 the school had been renamed as the Palmer School of Chiropractic and was successful enough to move to larger premises which became the nucleus of the twenty-first century campus, today occupying most of the Davenport district now named Palmer Hill.[23]

In 1906, while his father was serving a six-month prison sentence for practising osteopathy and medicine without a licence, B J staged a coup. He declared himself head of the Palmer School and his wife Mabel became its treasurer. 'Old Dad Chiro', as D D was by then known, was enraged. Once out of jail he opened a rival school round the corner from the first Palmer establishment and wrote a chiropractic textbook largely devoted to attacking his by now estranged son. As well as being a classic quack par excellence, Daniel David Palmer showed many signs of what modern psychiatry might describe as a narcissistic and dysfunctional personality. Delusional and grandiose, he fell out with everyone, including his own family and a total of five wives.[24]

D D Palmer died in 1913, three months after a road accident that occurred at the annual Lyceum and Homecoming, an event held every summer at the Palmer School of Chiropractic – now run by B J – which climaxed in a celebratory parade through the streets of Davenport. There are conflicting accounts of what occurred, but most witnesses agreed that 'Old Dad Chiro' turned up suddenly and uninvited, waving an American flag and insisting on leading the procession. An altercation followed, with B J Palmer arriving in his car to join the argument. 'What happened after that depends on whom you believe,' writes chiropractic historian Ralph Smith. 'Daniel David [Palmer] claimed that B J struck him with his automobile, and D D's friends and allies later produced affidavits of witnesses to prove it. B J flatly denied it, and produced many more affidavits to this effect than D D's cohorts were able to muster'.[25] Though the cause of Palmer's death was recorded as typhoid fever, there were those who thought it must have been precipitated by the injuries he had sustained at the parade. The legal wrangling continued long after D D's death.

B J Palmer would make his fortune through recruiting students to his chiropractic schools and courses and by promoting chiropractic around the world, particularly in Canada, Australia and the UK. He must have realised that the real money lay not in treating individual patients but in training chiropractors. He founded his school, he said, 'on a business, not a professional basis. We manufacture chiropractors. We teach them the idea and then we show them how to sell it'. Neither was he fussy about the quality of prospective students, telling one applicant 'in regard to educational qualifications, do not allow this to annoy you. We hold no entrance examinations'. B J had the walls of the Palmer School decorated with 'epigrams, admonitions, cartoons and witticisms', including the still-famous slogans 'the world is your cow – but you must do the milking', and 'early to bed and early to rise – work like hell and advertise'.[26] And his entrepreneurial spirit lives on today in those chiropractors who offer 'Free Spinal Checks' as a way of attracting prospective patients, as well as those who try to get patients to sign up for a never-ending course of treatment.

By the 1920s, the Palmer School in Davenport was thriving, with two thousand students and an annual income of $1,000,000. Many of the students were First World War veterans who had left the armed forces with no occupation to return to. B J started a mail-order business selling chiropractic equipment including adjustment tables, miniature spine sections, copies of his own portrait and the Neurocalometer, a heat-detecting device supposedly able to detect subluxations whether or not the patient had symptoms. Palmer held the patent and would allow it to be used only by graduates of his school. This controlled marketing of the Neurocalometer caused considerable friction in the chiropractic world, with three-quarters of the students at the school leaving in protest at what

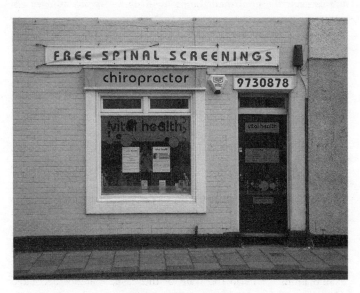

Chiropractic invitation

was felt to be Palmer's blatant money grabbing. It is still sold today as the Nervoscope, costing around $799 plus extra for accessories.[27]

In 1924 B J Palmer started the radio station WOC (Wonders of Chiropractic) 'the first radio station west of the Mississippi' and wrote thirty-eight books, including *Radio Salesmanship* and *The Bigness of the Fellow Within*.[28] Unlike his father, B J was a commercial and public relations genius, writes chiropractic historian Ralph Smith. 'The result was that Daniel David died penniless while B J, who took over the school and the leadership of chiropractic, died a multimillionaire' in 1961. In his 1926 work *Selling Yourself* he suggests that chiropractic patients should never be told that they are well, only that they are getting better. If a patient asks if chiropractic treatments are to be continued for life, the pat answer should be: 'No, only as long as you want to stay healthy'.[29]

Practice-building advice for the modern chiropractor comes from C J Mertz, former president of the International Chiropractors Association, through the scheme he calls 'Lifetime Family Wellness™'. He says 'ordinary practices become championship teams by mastering the fundamentals of a high volume, subluxation-based, cash-driven, Lifetime Family Wellness™ practice' and claims to have 'revolutionised 14,000 practices worldwide' through his lectures and franchise schemes.[30] One might hope that even the most unsuspecting client might be able to detect the hard sell implicit in the following outline written by Mertz which appeared in the *Chiropractor's Journal*, but it must be difficult to resist, especially while lying prone and semi-naked. This is how C J Mertz suggests chiropractors promote what he calls a Waiting List Practice:

'Step One: When talking to a prospective new patient your [assistant] asks "Is this appointment for you or your entire family?"

Step Two: During your pre-consultation, let the patient know you make a point of creating a family record for the detection of vertebral subluxation.

Step Three: In the examination, check the patient's posture in front of a mirror then give him or her a WLP family posture card to take home to survey family members.

Step Four: When reviewing X-rays in your report of findings, compare results from patient's posture card to encourage the patient toward a family check up, which is at your expense within the first week of his or her care.

Step Five: During the report of findings on second visit, have a beautiful registration book for workshop sign-up with automatic invitation for guest registration.

Step Six: Use 3x5 cards to develop new patient leads from workshop participants, then staple it to their travel cards and talk to them during office visits about ways to introduce those people to your practice.

Step Seven: Have patients write a testimonial as soon as possible regarding the success of their care – which is well known to cause patients to refer again.

Step Eight: Put hot seats near the adjustment room (door open) so patients can hear you talking about chiropractic and they can share stories among themselves which greatly inspires referrals.

Step Nine: Give a "WOW" gift for each patient's first referral which creates such gratitude, he or she feels compelled to refer again (even if you never give out another gift).

Step Ten: When a patient prompts the name of a prospective new patient, use a personalised [practice] referral packet in which to write that person's name and have your patients deliver it themselves.'[31]

In 2007 there were nineteen colleges of chiropractic in the USA. In the UK today the Anglo European College of Chiropractic in Bournemouth is one of two colleges offering accredited degrees in chiropractic. The college has around four hundred and fifty full-time students and claims to have trained over 80 per cent of chiropractors currently practising in Europe. Entry requirements are low, just three A levels at grade C or above in any subject. Fees are the same as at other universities in England and Wales: £3,000 per year. Students gain practical experience in what is described as 'the busiest chiropractic clinic in the world', claimed to see more than forty thousand patients every year at £20–£50 per visit, depending on the qualification of the practitioner. X-rays cost extra.[32]

The subluxation myth

As a single cause of all diseases – well, at least 95 per cent, according to Old Dad Chiro – chiropractic's vertebral subluxation has no equal in alternative or any other medicine. Not only that, it is believed to be a disorder present throughout life, as the first subluxation is said often to occur during birth.

Chiropractors say a subluxation 'is when one or more of the bones of your spine (vertebrae) move out of position and create pressure on, or irritate spinal nerves. Spinal nerves are the nerves that come out from between each of the bones in your spine. This pressure or irritation on the nerves then causes those nerves to malfunction and interfere with the signals travelling over those nerves'. This is Vertebral Subluxation Complex, and you would be forgiven for thinking it extremely serious because, as one information leaflet has it, 'if you interfere with the signals travelling over nerves, parts of your body will not get the proper nerve messages and will not be able to function at 100% of their innate abilities. In other words, some part of your body will not be working properly'.[33] Followers of D D Palmer's chiropractic philosophy are likely to reject germ theory and to believe that almost all diseases are caused by spinal irregularities. It is therefore no surprise to find that chiropractors might see no benefit in vaccination. One 2004 study of chiropractors in Alberta, Canada found that around a quarter of chiropractors actively advise parents against having their children immunised.[34] These findings have been confirmed elsewhere.

Most chiropractors believe in subluxations. The American Association of Chiropractic Colleges, an umbrella organisation for chiropractic teaching institutions, says they can be 'evaluated, diagnosed, and managed through the use of chiropractic procedures'.[35] The World Federation of Chiropractic reports that 65 per cent of its members surveyed believe the phrase 'management of vertebral subluxation and its impact on general health' fits chiropractic 'perfectly' or 'almost perfectly'.[36] The World Chiropractic Alliance, in pursuit of its 'vision of worldwide wellness' is 'dedicated to promoting a subluxation-free world' achievable only through a lifetime of visits to the chiropractor. You may be foolish enough to think yourself well because you have no symptoms of illness, but be warned: 'if a subluxation continues undetected, the organ may suffer great damage, yet we won't even be aware of it! Because subluxations can destroy a person's health without any visible signs, they are often called "the silent killer". So who needs this treatment? *Everyone* needs to make sure their bodies are free from subluxation ... That's why we *ALL* need regular chiropractic visits'.[37] A WCA survey found that 65 per cent of its members used the words 'silent killer' in their literature.[38]

Given its alleged ubiquity one might assume that the dangerous Vertebral Subluxation Complex would be easy to demonstrate. But this is precisely where chiropractic lurches headlong into pseudoscience and D D Palmer's concept of a 'religion'. These 'silent killer' subluxations are not only silent, but also invisible. There is simply no evidence they exist. No one has ever been able to provide any proof of the chiropractic subluxation, either through diagnostic tools such as X-ray, MRI or by examining the spine in a post-mortem examination. When Quackwatch's Dr Stephen Barrett challenged the Palmer College of Chiropractic to provide X-ray evidence of a subluxation, the school demurred, telling him 'chiropractors do not make the claim to be able to read a specific subluxation from an X-ray film. [They] can read spinal distortion, which indicates the possible presence of a subluxation and can confirm the actual presence of a subluxation by other physical findings.'[39]

Recently a small group of dissenting chiropractors, in a paper entitled 'Chiropractic – dogma or science?' has described the subluxation as 'an interesting notion without validation'. They say 'as it has throughout the past century, D D Palmer's mediating variable remains a "bone of contention" between many chiropractors and the scientific community, as well as among chiropractors themselves ... The chiropractic subluxation continues to have as much or more political than scientific meaning'.[40] Nevertheless, it appears that the majority of chiropractors not only believe in the existence of the subluxation but also think that most diseases are caused by spinal misalignment – more than 68 per cent of chiropractors in a 2002 Canadian study.[41]

John Badanes trained as a chiropractor in the US and was in practice for seven years until he realised that his profession had no logical or scientific basis. In an American television documentary he showed how easy it was for both chiropractor and patient to persuade themselves that the technique worked by, for example, demonstrating a difference in leg length by arranging the legs in a certain way before and after chiropractic adjustment. 'It's my position that they're all self-referential. In other words, they define a particular problem that incidentally has never been verified to be a problem, and then they have their methods of locating that problem, and then they have their specific methods of fixing it. So it's like a closed loop, and so it always works because you're just talking to yourself'.[42]

Or to put it another way, as does another US chiropractor, David Seaman, 'for nearly every chiropractor, there is a different description of subluxation and equally different methods of correction. Relativism in chiropractic, that is chiropractic relativism, also permits each chiropractor to define the nature of chiropractic in any way that he/she chooses'. Seaman still thinks it's possible, as he puts it, to 'build a chiropractic profession that is guided by real philosophy and real science'. But he must be aware that he has got his work cut out since he adds 'chiropractors are notorious for making treatment claims about chiropractic care that go well beyond the limits of our supportive data, whereas other professionals do not. Consequently, it is the chiropractor who looks like, and subsequently deserves to be called, an amateurish, unscientific huckster.'[43]

As long ago as 1963 US chiropractor Samuel Homola wrote a book supporting manipulation for back pain but rejecting what he described as chiropractic dogma. When his *Inside Chiropractic: a Patient's Guide* was published in 1999 he was called a heretic for describing chiropractic as 'a curious mixture of science and pseudoscience, sense and nonsense'. He confirms that the chiropractic profession has little tolerance for dissension. 'Its nonsense remains unchallenged by its leaders and has not been denounced in its journals. Although progress has been made, the profession still has one foot lightly planted in science and the other firmly rooted in cultism'.[44]

The X factor

It is generally known that an X-ray examination marginally increases the risk of contracting cancer. Because an X-ray involves exposing the body to radiation it should only be carried out when the benefit, say in order to make an accurate diagnosis, outweighs the risks.

One reason chiropractors require patients to have X-rays must be because they confer the respectability and status of a practice associated with scientific medicine. I counted thirty X-ray images in the 2006 prospectus for the Anglo European College of Chiropractic. It is not known how frequently chiropractors X-ray their patients and there appear to be no agreed clinical indications of when and whether an X-ray should be done. One study of European chiropractors found that in Italy 93 per cent of all chiropractic patients were X-rayed but in Sweden the figure was only 25 per cent. In the US a survey found that 96 per cent of new patients and 80 per cent of patients at follow-up visits were X-rayed.[45] Research in 2004 by the UK General Chiropractic Council showed the use of X-rays to vary wildly between practitioners, with some saying they use it for only a minority of patients and others who do it so often they must be X-raying almost anyone who walks through the door.[46]

Recent research into X-ray risks suggests that an average lifetime exposure to medical X-ray tests increases a UK person's cumulative risk of developing cancer by the age of seventy-five by 0.6 per cent. This sounds negligible but probably accounts for around seven hundred of the 124,000 new cases of cancer diagnosed in the UK every year. In the US X-rays are thought to cause one cancer in every 1,000.[47] Whilst there is no avoiding the low level of natural background radioactivity from our own bodies and from the environment, the UK Radiological Protection Board has set out how this compares with the amount of radiation given by an X-ray's dose of radiation: a dental X-ray amounts to a few days' worth of background radiation; a chest X-ray is the equivalent of ten days' worth; a spinal X-ray is the equivalent of six to twelve months of natural background exposure. The additional lifetime risk of cancer per chiropractic spinal X-ray therefore amounts to being something between one in 10,000 to one in 100,000.[48]

It is hard to see the benefit of chiropractic X-ray, especially when used in pursuit of non-existent vertebral subluxations or when it leads to the promotion of potentially dangerous adjustments. As Kurt Butler, in his *Consumer's Guide to Alternative Medicine* says, 'no one knows how many cases of cancer are caused by unnecessary radiation by chiropractors, but any number is too many because there is no benefit to justify any risk. The X-ray in the hands of some chiropractors is like a horoscope in the hands of an astrologer, but at least astrologers don't dose clients with radiation'.[49]

Chiropractic stroke

When 28-year-old Gloucestershire insurance negotiator Frances Denoon got a stiff neck after exercising, her GP told her to rest and take painkillers. But she wanted to speed her recovery in order to get back to work, so with her doctor's approval, consulted a chiropractor. At her second visit the chiropractor adjusted her neck with what Ms Denoon described to me as a 'gut-wrenching crack'. She was immediately nauseous and could hardly speak. 'I felt funny and couldn't focus and realised I wasn't saying the words clearly.' At first the chiropractor was not unduly concerned, but when her condition worsened he summoned a doctor from the health centre next door who sent Frances straight to hospital. As the ambulance doors closed Frances remembers her chiropractor's parting words: 'he told me that I would be alright now and it was a one in a million chance'. Only later did she realise that the 'one in a million' referred to what he must have known was a stroke.

Once in hospital it became apparent that Frances Denoon was suffering a severe stroke and swelling of the brain. She fell into a coma and needed three hours of neurosurgery to save her life. Eight weeks later she left hospital after making a partial recovery. In the years since the incident she has recovered enough to start a family but still has a number of residual effects including coordination and balance problems, as well as difficulty using her right hand for fine movements, such as writing. She told me she has only regained partial fitness and could no longer cope with the demands of her previous job.

The stroke was caused by the severing of one of the two vertebral arteries that supplied blood to Frances' brain. These arteries enter the skull alongside the topmost bones of the spine and if one is damaged it may swell and restrict blood flow. If it is severed a clot may form as it heals. If the clot becomes detached from the arterial surface it can block the supply of blood to the brain and thereby cause a stroke.

It was obvious to Frances Denoon as she began to recuperate that her injury and stroke had been caused by the chiropractor's neck manipulation. The medical team looking after her agreed, indeed her surgeon told her he believed it could only have been caused by the chiropractor's forceful twist of her neck. This move, which arguably amounts to the hallmark of chiropractic manipulation, involves extending the neck and then making an extreme and brisk rotation of the head. It is used principally to treat symptoms such as a stiff neck or headaches but is frequently employed in the treatment of other problems, including lower back pain and even illnesses such as asthma. Vertebral Arterial Dissection – tearing of the vertebral arteries – is a recognised risk of this manipulation. There is no clear picture of the incidence of VAD following treatment by a chiropractor, but there are many worrying reports of the problem.

Frances Denoon took her chiropractor to court but lost the case because she could not demonstrate negligence. Although there was no question that her vertebral artery had sustained damage, the fact that the chiropractor had performed a standard chiropractic adjustment in the approved manner meant there was no case for him to answer. He was just doing what chiropractors do and he was doing it correctly. The experience led her to found the campaigning group Action for Victims of Chiropractic, in order to publicise the risks.[50] She has been inspired by the work of an American campaigning group, Neck911USA, which has used bus advertising to get the message across.

In Canada in 1996 Lana Dale Lewis died at the age of forty-five, following a stroke which occurred ten days after a chiropractic adjustment. She had seen a chiropractor for help with migraine but suffered immediate pain and dizziness after her first neck manipulation. After the second manipulation she died. The inquest into her death lasted two and a half years and ruled accidental death, not natural causes. The coroner's jury was legally prevented from assessing fault or blame, but was advised that a verdict of accidental death meant they attributed the death to the neck manipulation.[51]

Then two years later a 20-year-old woman died a day after chiropractic treatment for pain in her lower back. Within minutes of a neck manipulation Laurie Jean Mathiason of Saskatoon, Saskatchewan was having convulsions and was rushed to hospital in a coma. Doctors declared her brain-dead the next day. Afterwards Laurie's mother and fiancé, who were present at the time, said the only thing the chiropractor did to try to revive Laurie Jean was to slap her face.[52] Giving evidence at the inquest, a radiologist testified that Laurie Jean's vertebral 'blood vessel wall was severely damaged, torn. It was the direct result of a forceful manipulation to the neck. This was an acute blood clot'. The cause of death was found to be 'traumatic rupture of the left vertebral artery'. As in the case of Lana Dale Lewis, the coroner's jury was not permitted to attribute blame, but the pathologist who performed the post-mortem on Laurie Jean Mathiason testified that he thought there was a 99 per cent likelihood that chiropractic neck manipulation was the cause of death.

Edmonton neurologist Dr Brad Stewart helped care for Laurie Jean Mathiason. Her case was a turning point, he said in an Irish television report. 'I got tired of seeing people hurt, and I didn't like seeing a twenty-year-old girl die because of lower back pain. This wasn't something I was seeing once every six months or once a year – we were seeing strokes every month at the University Hospital. Every month we'd have yet another person come in [after chiropractic neck manipulation] with a dissection in the neck and a small stroke or a large stroke. The attempts to minimise this by the chiropractic profession are astonishing and to be frank rather disgusting'.[53]

Dr Stewart led a group of sixty-two colleagues to issue a warning about the dangers of neck manipulation, describing it as one of the leading causes of stroke in the under forty-fives. 'The public must be made aware that the neurological damage that can result subsequent to upper neck manipulation can be debilitating and fatal,' they said. They joined with Canadian paediatricians in calling for a stop to chiropractic treatment of children and babies for conditions like colic, ear infections and bedwetting, for which there is no evidence of benefit and which only 'expose infants and children to unwarranted neck manipulation'.[54] The government of Ontario has now withdrawn public funding for chiropractors.[55]

Research into VAD and stroke after chiropractic manipulation has produced contradictory results, so its real prevalence remains unquantified. Chiropractors put the risk at one in every million neck adjustments, but other reports indicate the incidence could be much higher. In 2002 Canadian researchers found that stroke victims were five times more likely than controls to have visited a chiropractor in the week before their stroke.[56] Other Canadian neurologists have estimated that as many as 20 per cent of strokes in the under forty-fives are due to neck manipulation.[57] A German account published in 2006 described thirty-six patients who experienced VAD after neck manipulation.[58]

In an article reviewing a number of reports on the safety of spinal manipulation, chiropractic sceptic Professor Edzard Ernst suggests there is convincing evidence to demonstrate that 'neck manipulation is associated with frequent mild adverse effects as well as with serious complications of unknown incidence'. Mild effects follow around 50 per cent of neck manipulations and there are reports of other consequences including epidural haematoma, intracranial aneurysm and peripheral nerve palsy. A recent American review of adverse effects of spinal manipulation suffered by children described fourteen cases gathered from eight electronic databases: nine involved serious adverse events (subarachnoid haemorrhage, paraplegia), two involved moderately adverse events that required medical attention (severe headache), and three involved minor adverse events (back soreness). Another twenty cases involved delayed diagnosis (diabetes, neuroblastoma) and/or inappropriate provision of spinal manipulation for serious medical conditions (such as meningitis and rhabdomyosarcoma).[59] Despite such disturbing reports, says Professor Ernst, 'one gets the impression that the risks of spinal manipulation are being played down, particularly by chiropractors'.[60]

In Britain the General Chiropractic Council has warned against what they call 'alarmist misinformation' about chiropractic and strokes. They are confident that the incidence of stroke after neck manipulation 'is no more than would occur naturally within the general population. It is a sad fact that strokes occur after many common events – those who are predisposed to suffer a stroke could do so as a result of everyday activities like turning to back their car or having a violent bout of coughing'.[61]

In a 2005 television interview Sharon Mathiason, mother of Laurie Jean, recalled what one doctor said as she accompanied her dying daughter to accident and emergency:'Never let those buggers touch you above the shoulders'.

No touch non sense

There is one form of chiropractic that definitely does not injure the neck. It is known as NUCCA and is a method practised by members of the National Upper Cervical Chiropractic Association. NUCCA was invented in the 1940s by two chiropractors, Ralph Gregory and John Grostic. It is, in essence, no-touch chiropractic. As one practitioner says 'The NUCCA Spinal Correction is unique in the field of chiropractic, designed to restore body balance and normalize the flow of healing messages from the brain to all parts of the body so it can reactivate its natural self-healing process. The NUCCA spinal correction requires NO CRACKING, TWISTING OR POPPING OF THE NECK OR BACK' (*sic*).[62]

Harriet Hall, a retired doctor and campaigning sceptic known as the SkepDoc, came across an internet video demonstration of the technique by Seattle-based NUCCA practitioner Johanna Hoeller.[63] In the video Hoeller 'adjusts' a patient without touching them. Instead she cracks her own wrist in close proximity to the patient's neck. Dr Hall thought it was abundantly clear from the video 'that Hoeller believed she is touching the patient when she was not and if she was not lying to defraud patients, then she is clearly delusional' and complained to the Washington State Chiropractic Board and to the State Department of Health. But her reports have not been acted upon because no patient has complained about Johanna Hoeller. This led Dr Hall to enquire whether 'I may legally succumb to a similar delusion without fear of losing my medical license; that I may charge patients for pretending to perform surgery without touching them, just as long as no patient complains'.[64]

Head case – craniosacral therapy

Craniosacral therapy (aka cranial osteopathy) was created in the 1930s by an American, William Sutherland. A student of Andrew Still's School of Osteopathy he went on to 'discover' that the bones of the skull were in continual motion in response to a hitherto unknown pulsation of the brain he named 'craniosacral rhythm'. Sutherland believed the cerebrospinal fluid that surrounds the brain and spinal cord contains what he called the 'Breath of Life', a term he gleaned from the Bible, which consists of 'a potency, an intelligent potency, that is more intelligent than your own human mentality'. Its nature is much the same as Chinese medicine's Qi or homeopathy's *vis vitalis*. The Breath of Life is supposedly enhanced when the hands of the practitioner exert a pressure no greater than five grams (the weight of a small coin) on the head and back. This is intended to correct the flow and balance of the craniosacral rhythm and thereby maintain health.[65]

Craniosacral therapy's major logical fissure is a simple matter of anatomy. The bones of the skull are fused by the time we reach late adolescence and cannot be manipulated in adulthood. Neither is there any objective evidence for a craniosacral rhythm beyond the pulse produced by the heart.[66] John E Upledger, described in promotional material for his autobiography as 'the man who had the courage to see beyond the limits of what science said was even possible – and the nerve to take on the medical establishment', is currently the face of craniosacral therapy in the US. He claims to have proved the bones of the adult skull to be moveable, but these findings have yet to be replicated.[67] But these facts don't stop the Craniosacral Association of the UK claiming the therapy 'has a very high success rate. Most practitioners find that around eighty five per cent of clients are happy with the results of the work.'[68] Eighty five per cent? Well I never. As we have seen repeatedly, one of the perennial signs of the quack is their claimed success rate lurking somewhere around the eightieth centile.

What really gives the game away is that when the 'craniosacral rhythms' of the same patients are measured by different therapists there is no agreement beyond that expected by chance. As one research paper puts it, 'our own and previously published findings suggest that the proposed mechanism for cranial osteopathy is invalid and that interexaminer (and therefore, diagnostic) reliability is approximately zero'.[69] Treatment by a craniosacral practitioner is therefore entirely based on the subjective response of the therapist. My favourite study shows that the therapists' measurements of cranial rhythm were more likely to be determined by their own breathing rates than by any other factor.[70] Practitioners would no doubt say that this indicates how tuned in they are to their patients, since they do seem to have an answer for everything. For instance they already manage to rationalise those pesky fused skull bones by saying 'we consider the bones of the skull to be a very dense *fluid*'.[71]

There is, though, one section of the population that is definitely ripe for manipulation – the sleep-deprived parents of newborn babies. As Brid Hehir of Camden NHS Primary Care Trust says, 'it is not surprising that some parents are attracted to CST. Bringing up a baby can be an anxious time, and when your baby develops something like colic, which can seem impossible to get rid of, a therapy that claims to target the supposed core of our entire physical wellbeing can seem attractive'.[72] Calculatedly or not, CST therapists manage to capitalise upon such anxiety. One North London clinic concentrates the worried parental mind with the claim that 'craniosacral therapy is uniquely effective in dealing with the effects of Birth Trauma (both the compressive forces exerted on the skull, and also the shock effects of a traumatic birth), with its many possible consequences including dyslexia, learning difficulties, hyper-activity, behavioural disorders, epilepsy, autism, asthma, squint and cerebral palsy'.[73] They usually offer free check-ups for infants under six months old, after which they charge upwards of £30 per session, sessions which can be performed, conveniently, when the baby is asleep. The following sales patter, from Tenterden-based therapist Bill Ferguson, is typical: 'the nature of your birth determines the nature of your cranium and consequently influences the way you develop (personality, intelligence, memory etc) . . . Birth trauma can be treated all through life but the sooner it is treated, the faster and more complete the response'.[74] To the craniosacral therapist's mind, all births are traumatic, yet another example of the universal diagnosis so popular in alternative medicine.

'Birth trauma' can, we are told, have a negative impact on the skull of any baby, even one born by Caesarean section (this time due to the sudden change in pressure) and is alleged to lead to problems with feeding or settling, as well as colic and crying. No shortage of prospective customers there, and all with problems that are known to cease with time. Infant colic in particular almost always resolves when the baby is between two and three months old – hence 'infant' colic – so it's easy for a therapist to take the credit when it stops. Like London therapist Nicole Ragueneau who 'sees a lot of babies under three months' and who told *The Times* she is able 'to listen to the body's intrinsic movement patterns in order to diagnose and facilitate with therapeutic procedures the release of any congestion and resistance'.[75] It's also called wind.

Brid Hehir of Camden Primary Healthcare Trust says that midwives, GPs and health visitors now routinely promote craniosacral therapy, granting it what she calls an 'undeserved legitimacy'. This is especially concerning since CST therapists in the UK don't have to be registered or to have had any previous health-care experience, not even in osteopathy or chiropractic. Courses are open to anyone who can stump up the fees. The UK Institute of Craniosacral Studies offers a two-year training course, spread over sixteen weekends, at a cost of £6,600. Applicants will also need to pay for their own treatment, since 'due to the process-oriented nature of the work, students are required to receive regular Craniosacral Therapy sessions throughout the duration of the course'. By the end of the two years they will have developed 'a direct sensory awareness of their own systems' and set up as a craniosacral therapist. In preparation for their diminutive clients, there are two whole days devoted to embryology.[76] A shorter, cheaper course – eight weekends over eight months for just £945 – is offered by the Craniosacral Therapy Educational Trust.[77]

Touch typing

Therapeutic touch is where modern manipulative therapies like osteopathy and chiropractic meet mesmerism. New York nurse and theosophist Dolores Krieger decided in the early 1970s that she could assist natural healing processes 'by redirecting and rebalancing the body's energy field'. She believed the palms of the hands to be chakras that can transfer healing energy to the patient by making sweeping passes a few inches away from the body.[78] It's the laying on of hands without the laying on, faith healing without the faith.

In all fairness it is more complicated than that, because the TT process has a total of four 'dynamic and interactive phases'. First of all the therapist 'centres' the patient by 'bringing the body, mind, emotion to a quiet focused state of consciousness'. Then she assesses the patient 'by holding the hands two to six inches away from the individual's energy field while moving the hands from the head to the feet'. How do they decide exactly how many inches? In any case, 'assessment can be done in a matter of minutes'.[79] Then comes 'intervention' which is TT's central feature and comes in the form of what is called 'clearing' or 'unruffling'. This is supposed to facilitate 'the symmetrical flow of energy through the field'. 'Unruffling' is achieved 'by using hand movements from the midline while continuing to move in a rhythmical and symmetrical manner from the head to the feet'. Then, following a little 'balancing' and 'rebalancing' to recharge the energy field and unblock 'areas of congestion' the therapist proceeds to a state of 'evaluation/closure', in which they use 'professional, informed and intuitive judgement to determine when to end the session'.[80] Or maybe they just look at the clock.

It is all so very silly sounding – especially the 'unruffling' part – that you won't be a bit surprised to learn there is no scientific evidence in support of Therapeutic Touch or for the existence of a human energy field. Indeed in 1998 such claims were comprehensively trashed by ten-year-old Emily Rosa, who in doing so became the youngest person ever to have a paper accepted by the *Journal of the American Medical Association*. Following a study at a school science fair for which she won a blue ribbon, she tested twenty-one TT practitioners to see if they could detect her aura from behind a cardboard screen. Each subject was tested between ten and twenty times. The practitioners were asked to place both hands through holes in the screen and rest them on a tabletop, palms up. Emily flipped a coin to decide which of the practitioner's hands she would place hers near. She would hover her hand, palm down, a few inches over one of the subject's hands. The practitioners then had to say where her hand was and whether or not they could detect her 'energy field'. They were able to get it right about half the time, a rate no better than what would be expected by chance.[81] She went on to win a 'Skeptic of the Year' award and an Ig Nobel Prize from the journal *Annals of Improbable Research*.

Despite such findings Therapeutic Touch is still popular and widely practised, both by alternative practitioners and mainstream health-care professionals, especially American and Australian nurses. One US sceptic believes that nurses should never be permitted to use TT and it could be more accurately renamed CPEH – Close Proximity Empathetic Hovering.[82] In common with so many alternative practitioners, they invoke quantum theory to support their claims, even managing to implicate an innocent and unknowing Albert Einstein: 'the underlying principles upon which this technique is based include acceptance of the Einstein paradigm of a complex energetic field-like universe.'[83] In this case it's a universe constituted of 100 per cent hot air.

Chapter 7

Cancer – when it really matters that alternative medicine doesn't work

Prostate specific antigen (PSA) is a natural chemical produced by the prostate gland in men. A rise in levels of PSA can be an early sign of prostate cancer, so when a 75-year-old Australian we will call Peter Johnson tested positive for raised levels of PSA, he was worried.

Rather than pursuing the medical investigations that would have confirmed whether or not he had prostate cancer, Johnson set about researching the condition on the internet. He would have found as many as 287,000 sites suggesting that selenium may have a role in both the prevention and treatment of prostate cancer. This trace element is found in many foods, including seafood, grains and eggs, but it is also available as a nutritional supplement said to help protect tissues from attack by damaging free radicals. Selenium is one of the cheaper supplements – it costs around £8.95 for 360 tablets from UK health supplement stores. But Peter Johnson mistakenly purchased 200 grams of sodium selenite powder, normally used as a supplement for livestock. As there were no instructions for how much to take, he decided to swallow a dose of 10 grams. He had no way of knowing that this was approximately ten thousand times the recommended supplementary human dose.

Three and a half hours later Johnson was in Brisbane Hospital suffering abdominal pain, low blood pressure and a number of other signs of selenium poisoning. He was taken immediately into intensive care where he suffered a cardiac arrest from which he could not be resuscitated. Just six hours after swallowing the selenium overdose, Johnson was dead. The medical team who had tried to save him decided to alert colleagues by writing up the case for the *Medical Journal of Australia*, concluding that it 'highlights the risks associated with failure to critically evaluate internet material and exposes the myth that natural therapies are inherently safe'.[1]

Cancer patients use complementary and alternative medicine more than anyone else. Before the late 1990s only around 20 per cent of UK cancer patients used it but surveys show that use has now increased, in parallel with the general rise of CAM across the wider population: current estimates range between 31 and 80 per cent of all cancer patients. Use is also often related to the kind of cancer a person has – CAM is used by just 24 per cent of brain tumour patients but by as many as 75 per cent of women with breast cancer.[2][3]

The fears which inevitably accompany a potentially life-threatening illness make us easy prey to the false hope offered by some CAM scammers. Here I will expose several obvious fraudsters who have duped dying people of their savings in return for a spurious 'cure'. But this story is more complex than one of wicked fraudsters and innocent victims; CAM has spread throughout the world of cancer care and as it spreads it reveals much about contemporary attitudes to illness and to science. The encounter between cancer and CAM exposes the notion that we are individually responsible for our own health outcomes; it highlights our tendency to blame the victim and it sees the alternative health critique of scientific medicine at its most extreme. Above all, it generates a fatal relativism that says whatever a person might choose to do to defeat their cancer is acceptable because all therapies – orthodox and alternative – are equally valid.

Although most cancer patients who use CAM use it alongside orthodox treatment – surgery, chemotherapy, radiotherapy and the like – a very small minority reject all mainstream approaches and use only alternative methods. Some of these people may go on to use orthodox medicine later on, if and when the disease progresses. This delay will almost certainly mean the cancer is more resistant to treatment. Unfortunately there is little research or follow up of those who exclusively use alternative medicine as they are likely to detach themselves from mainstream medicine and so be difficult to contact. What the few small existing studies suggest is that progression of the disease is frequent in those who eschew orthodox treatment.[4] [5] And one eight-year study of 515 cancer patients found that death rates were actually higher amongst those who used alternative medicine alongside orthodox treatments, concluding that 'the use of [alternative medicine] seems to predict a shorter survival from cancer'. This effect appeared mainly in those who initially had a good prognosis.[6]

There are now reports of cases of advanced, untreated cancer producing catastrophic symptoms that doctors have not encountered for decades. For retired Australian surgeon Peter Moran, who has written extensively about alternative medicine and cancer, the 'concern is not merely the as yet rather small number of lives lost unnecessarily, it is the foregoing of the very substantial palliative benefits that conventional methods can offer . . . it is appalling to me, for example, that [today] some breast cancers are being allowed to progress to cancer-en-cuirass (widespread cancer of the upper body) and other miserable states, scarcely otherwise seen for nearly a century.'[7]

Spoilt for choice

The idea that one can and should take responsibility for one's health is most powerful in relation to cancer. This can lead cancer sufferers and their families and friends to feel obliged to try any alternative therapy promising to reduce distress and to improve their chances, regardless of the contradictions between the many CAM philosophies and practices. Whatever a person might choose to do when they have cancer, from taking dubious herbal remedies to going for a short walk, gets defined as a 'complementary' therapy and is deemed beyond question because the cancer patient herself has chosen it.

Cancer is a horrible disease that usually necessitates unpleasant and lengthy treatment. It kills. One in three of us will get it, mostly when we are older. We all have an urge unconditionally to support those with cancer who want to improve their situation. It's a kind of 'whatever gets you through the night's all right' acceptance of cancer patients' entitlement to make their own decisions about what they do to manage, both emotionally and physically, the illness and its onerous treatment.[8] Faced with the potential ravages of cancer, few would want to be denied at least the semblance of control. And whilst this is perhaps an explanation, it is certainly not an excuse for the extreme relativism amongst both doctors and cancer sufferers about the use of alternative therapies. Doctors do not want to appear closed-minded. And because they are anxious about upsetting and antagonising patients, they are all too often unwilling to say outright that a complementary therapy is plain useless. It is this relativism that is so often exploited.

As the UK organisation Cancer Options, who offers 'counselling, consultancy and training' for cancer sufferers, affirms: 'there are no rights and wrongs about how you deal with your cancer, just what is the right thing for you at the right time. If you need the powerful effects of orthodox treatments, that is fine, if you need a more gentle, holistic approach, that is also fine'.[9] Those who seek out alternative therapies for cancer are likely to be especially vulnerable to such dangerous and misleading statements. One research study involving 480 women with breast cancer found that those who turned to CAM may be unusually worried and depressed and turn to alternative medicine principally for help in coping with the stress of their illness. Women using CAM reported a lower quality of life, a greater incidence of depression, more fear of recurrence of cancer and lower sexual satisfaction.[10] [11]

Consciously or not, CAM providers often use language that suggests a value for complementary therapies for which there is no medical evidence. So a woman with breast cancer might reasonably conclude that her chance of healthy survival could be improved by attending the Breast Cancer Havens in London and Hereford, for example, since they describe themselves as 'places specifically designed to provide a safe environment where mind and body can start to heal'. The Haven (Patron HRH the Prince of Wales) makes no claims of cure beyond this generalised notion of 'healing'; therapies director Caroline Hoffman told me that the role of the Breast Cancer Haven is essentially to 'help to hold people in a safe place while they're going through this traumatic experience'.[12] But this appears at odds with founder Sara Davenport's description of the

Haven as 'a place where miracles can happen'.[13] There is no question that the help provided at Breast Cancer Haven is motivated by altruism and genuine concern; that one of the principal intentions is to support those undergoing orthodox treatment and that some people find this useful. Their promotional material is full of testimonials from women who say they have found the place to be calm, tranquil and helpful at a time of great need.

But the Breast Cancer Haven also offers a range of therapies described as if they supply something beyond emotional sustenance. As well as free counselling, exercise classes, relaxation sessions and group support – all of which have been shown to make a real improvement in the quality of life for people with cancer – Breast Cancer Haven provides (among other things) homeopathy, craniosacral therapy, Emotional Freedom Technique (they call it a 'gentle, highly effective treatment tool'), Indian head massage, kinesiology, reflexology, reiki and shiatsu. They also promote NES Energetic Analysis, using a machine that 'can effectively map the human body field, establishing which organs are lacking in energy, where blockages have been triggered by environmental chemicals'.[14] But the NES system relies on the same junk science as the Quantum Xxroid Consciousness Interface laptop we have already encountered. Its inventors claim that when a client places her hand on the 'input device' something called 'quantum entanglement' takes place and 'this information is carried via subatomic particles to the computer'. NES also sells products they call 'info-ceuticals', which they've created through 'a proprietary method of imprinting Quantum Electro-Dynamic body field information onto a base of organic micro-minerals.'[15]

Pity the woman with breast cancer, trying to make sense of such foolishness in the midst of probably the most serious health crisis of her life. Some alternative establishments, such as the new £7 million Penny Brohn Cancer Care (formerly the Bristol Cancer Help Centre) are unabashed about their potentially misleading use of language. When they offer 'healing' they say 'in this context, healing does not specifically equate with the verb "to heal". This therapy does not claim to heal but to sustain, restore balance, simultaneously calming and lifting the spirits'.[16] At least we've got that straight. No actual healing, then.

As well as the altmed cancer institutions there is a profusion of books promoting alternative treatments for cancer. Dr Rosy Daniel is a well-known 'integrated practitioner' who has written an encyclopaedic *Cancer Directory* in which she presents a myriad of alternative cancer treatments as if they are all of equal value. In a chapter entitled 'Alternative Cancer Medicines – the best of today' she lists anti-cancer nutrients, herbal medicine, hormone therapy, metabolic therapy, immunotherapy, neuroendocrine therapy, physical therapy, nutritional therapies and mind-body medicine, but fails to define what she means by 'best' and does not set out any useful exposition of efficacy. Instead the choice of cancer treatment is once again presented as a matter of personal taste, rather like choosing a shampoo. 'While the number of options available may perhaps make it difficult to choose what is right for you, it will give you a good basis for discussion if you choose to visit any integrated or alternative cancer doctors or clinics,' writes Dr Daniel.[17] What can she mean by 'right for you' in this context? Is it the least unpleasant option, or the one that has been shown to be the most effective treatment or cure for your cancer? These vital questions go unanswered.

And what is the cancer patient to make of another of Dr Daniel's enterprises, the 'Health Creation Cancer Lifeline Kit'? Lifeline? Isn't a lifeline something that saves your life? 'Health creation'? Doesn't that sound like 'cure'? Her stylishly produced Cancer Lifeline package of books, CDs and videos, sells for £150. Attending a series of four 'Health Creation Programme' weekend workshops costs £1,500. The programme is designed 'for all those who want to learn how to make the maximum improvement in achieving health, happiness and peace of mind . . . in an atmosphere of positive support, with the inspirational guidance of its author Dr Rosy Daniel'.[18] Health Creation is advertised by testimonials such as this from 'Anthony' who was diagnosed with pancreatic cancer (which has a very poor prognosis) in 2003: 'I strongly believe by taking Health Creation's holistic approach has helped me become a disease free man. I feel fit and healthy and I'm looking forward to the next phase of my life.' But one man's convictions about the reasons for his survival, however sincere, do not amount to reliable evidence when making a decision about cancer treatment.

It doesn't have to be like this. It is possible to offer care, support and information for cancer patients without homeopathy, fanciful tales of quantum entanglement, blocked energy pathways and healing that doesn't heal. Following her own experience as a cancer patient in NHS hospitals, the architect Maggie Keswick Jencks inspired the foundation of a series of purpose-built and architecturally beautiful Maggie's Centres for cancer patients in Scotland. She wanted to provide a homelike environment which would 'encourage and not intimidate'.[19] Maggie's Centres aim to provide 'high quality, evidence-based psychological support' built alongside NHS hospital facilities in order to help patients gain the confidence to live life after cancer 'with hope and determination'.[20] Such a setting should be a basic entitlement for anyone dealing with the challenge of a serious illness and a central feature of all modern cancer treatment, underpinning a palliative approach that relieves symptoms without any need to suggest that it's curing cancer.[21] This was what motivated Dame Cicely Saunders' founding of the modern UK hospice movement in the 1950s. St Christopher's was the first hospice to bring together teaching and clinical research, pain and symptom control, and compassionate care. It was 'a place where patients could garden, write, talk – and get their hair done'.[22] It was never claimed that these activities might cure cancer or prolong life.

Maggie's Centre Dundee, designed by Frank Gehry

The big C: conspiracy

Cancer brings together the most duplicitous and dangerous manifestations of CAM culture and practice. Here we find the greatest hostility to orthodox medicine and the scientific method. All too often promoters of alternative medicine push the idea that medical science is next to useless in the treatment of cancer, despite statistics that show higher survival rates. Worse still, they accuse doctors, researchers and drug companies – known in altmed as 'The Cancer Industry' – of conspiring to suppress 'natural' treatments. The suggestion is that the big money is in promoting inadequate cancer therapies, because an effective cancer cure or preventive would do everyone out of a job. This charge is the most acute, dangerous and paranoid manifestation of the recurring belief in CAM: that the 'medical establishment' is actively trying to stifle alternative medicine in order to retain its power.

They say there's a lot at stake. One American alternative medicine practitioner and supplement salesman, Gary Null, tells us that 'a solution to cancer would mean the termination of research programs, the obsolescence of skills, the end of dreams of personal glory, triumph over cancer would dry up contributions to self-perpetuating charities . . . It would mortally threaten the present clinical establishments by rendering obsolete the expensive surgical, radiological and chemotherapeutic treatments in which so much money, training and equipment is invested . . . The new therapy [alternative medicine] must be disbelieved, denied, discouraged and disallowed at all costs, regardless of actual testing results, and preferably without any testing at all'.[23]

In an article titled 'Impossible Dream – Why The Cancer Industry Is Committed To *Not* Finding a Cure, and If They Do Run Across One, Suppressing It!' we discover how those who run altcancer.com believe the mind of a cancer researcher works. Their imaginary Cancer Industry researcher says 'every year we must show you results. After all, you won't support us if you don't think we're getting something done. On the other hand, we can't be *too* successful – and we certainly can't afford to come up with a cure – after all, if we did that, how could we come back to you next year and get more of your money?'[24] When in 2003 the US Food and Drugs Administration stopped Alpha

Omega Labs selling Cansema, a worthless cancer cure, one supporter suggested that this was 'no doubt because their products worked. The FDA has a long history of doing this to developers of successful cancer remedies'.[25] Far-fetched as these accusations may seem, they are believed by many. Research by the American Cancer Society found that 27 per cent of respondents agreed that the medical establishment was suppressing a cure for cancer and that a further 14 per cent thought it might be true.[26]

Alternative cancer therapists say their plant-based 'cures' are over-looked by pharmaceutical companies because naturally occurring substances – rhubarb, for example – can't be patented, precluding profit for 'Big Pharma'. But David Colquhoun, Professor of Pharmacology at University College London, disagrees. He told me 'the kudos that a pharmaceutical company would get for finding an effective cure for cancer would be so enormous that it's hard to imagine that they would decline to produce it, even if it didn't make a lot of money. In any case, even when a plant-based substance (like Taxol from yew) provides the initial lead, it is common for synthetic derivatives to be made that have better properties than the original. Also, it is common to get patents on the methods of isolating or administering the substance, even if the structure itself can't be patented'. In other words, even if you can't make money directly from rhubarb, there'll be plenty of profit and acclaim when you develop a safer, more effective – and patentable – rhubarb-related derivative.

Oncologists, in common with the rest of the medical profession, already prescribe a number of substances derived from plants and other natural sources. The chemotherapy drug Vincristine was derived from the rosy periwinkle and the skin cancer treatment Cantharidin originated from the Spanish fly beetle. Researchers are also investigating the possibility that aspirin derived from willow bark – way past any patent opportunity – might have a role in the prevention of cancer of the mouth, throat and oesophagus.[27] Overall about a quarter of all prescription drugs are taken directly from plants or are chemically modified versions of plant substances.[28] Certainly the haste of the drug companies to support the development of new plant-based medicines in China belies the alternativists' claims that 'Big Pharma' always seeks to deny the possible effectiveness of treatments found in nature. The recent collaboration, for example, between the German pharmaceutical company Merck and the Hong Kong firm Chi-Med is primarily aimed at developing new anti-cancer drugs.[29]

Both earlier diagnosis and research have led to a battery of new cancer treatments and the hope is that in future the disease will become a manageable condition, like diabetes, heart disease or asthma are today.[30] But cancer quacks love to focus on the rising total of cancer diagnoses and neglect the improvement in survival rates. Promoters of one treatment claim 'our cancer statistics are worse than ever. Is it money or politics that is blocking the cure?'[31] If your oncologist fails to support your use of alternative remedies they say it's because she is an agent of the Cancer Industry – or its equally dangerous cousin 'organised medicine' – and motivated by both greed and fear: 'Don't expect a doctor working inside the system to buck the system. The risks are too great'.[32]

Quack cancer treatments are described as 'the natural cures they don't want you to know about' and the mainstream oncology practices of surgery, radiotherapy and chemotherapy are characterised as 'slash, burn and poison'. Despite the fact that these treatments demonstrably do work for many people again and again in the quantifiable sense that they either extend life or even cure the cancer, the cancer quacks tell us 'you cannot *poison* your body into health with drugs, chemo or radiation. The holistic approach treats the whole animal, ignites the body's internal healing force and stimulates the body's natural abilities to heal itself'.[33]

The cancer field is also where many contemporary myths and misunderstandings of the effects of the 'Western' lifestyle reside, and where there are the most misleading renderings of that pernicious half-truth 'you are what you eat'. But curiously, alongside all the alarming discussions about cancer-causing chemicals in face cream, little is said about the plant tobacco as the leading cause of avoidable cancer, or that lung cancer is the biggest killer. And rarely is there any mention of the substantial evidence that exercise is of real benefit to cancer sufferers both during and after treatment. That exercise is a non product-related self-help activity, which costs nothing, is surely relevant.[34]

An assessment of thirty-two of the most popular websites on complementary and alternative medicines, published in *Annals of Oncology*, concluded that they offered 'information of extremely variable quality. Many endorse unproven therapies and some are downright dangerous'.[35] The internet is teeming with campaigners against orthodox cancer treatment, who more often than not base their arguments on a mixture of anecdote and something very like spiritual faith. Agnes Tiller, a webmaster on the internet website curezone, vehemently opposes any materialist physical view and appears determined to undermine mainstream medicine by frightening the cancer patient away from it. Tiller warns 'no healing method that focuses only on the "physical" body will ever have a significant curing rate of any cancer . . . my definition of CURE is that you are cured when your cancer or some other serious disease or serious accident will NEVER happen again. NO person, I REPEAT, NO person who only addresses the curing of the "physical" body will EVER reach this level of cure . . . those who addressed ONLY the health of the "physical" body, later experienced other serious life problems, either the same type of disease or being a victim of an accident'.[36]

Others declare that orthodox treatments are inherently dangerous. This statement, from the website cancerfightingstrategies, is typical of the altmed view: 'Chemotherapy and other treatments *damage cells* and tear down and weaken the immune system. But the problem in the first place is that your immune system is already weak, and that your cells are already damaged. Even if tumours do go into remission, these treatments will have damaged other cells, which are more likely to turn cancerous. The immune system, unless it is supported by supplements and diet to help it recover, will be in worse shape then ever. While it may have taken decades for cancer to develop the first time around, the second time usually takes a year or two'.[37] This is misleading and fallacious from start to finish. Cancer is not caused by a weakened immune system, which is able to recover fully after chemotherapy and can be supported by other drug treatments during treatment if necessary. Every treatment has its risks and benefits, but the specific benefit offered by chemotherapy – the killing of cancer cells – is one that in the vast majority of cases outweighs the risks.

Shark infested waters – some disproven alternative therapies for cancer

'Do you find complementary therapies confusing?' asks Canceractive's *Icon* magazine, a British publication that promises 'everything you need to know to help you beat cancer'. Anyone would be baffled after reading the magazine's own feature articles. These have included 'Living proof – curing melanoma with Gerson', 'Photodynamic therapy – the light fantastic' and 'Iridology – your life in your eyes'.[38]

Most CAM cancer treatments are supported by no more evidence than the customary patient testimonial. Many have been around for decades, like the Hoxsey herbal tonic, banned for sale in the USA since 1960 but still readily available via the internet and direct from a clinic in Mexico.[39] It claims the mandatory 80 per cent success rate. The herbal cocktail Carctol is a comparatively recent invention, created by an ayurvedic practitioner in India, Dr Nandlal Tiwari, and available in the UK only on private prescription from a handful of 'integrated practitioners'. Widespread publicity about this product a few years ago, from Dr Rosy Daniel and others, has not led to any published evidence of effectiveness, despite it being heralded as 'a formula that gave new life'.[40] Or there's Cancell, invented by chemist Jim Sheridan after the idea came to him in a dream some sixty years ago. Sheridan also claimed that Cancell has an 80 per cent cure rate (yes!) for cancer, diabetes, arthritis, lupus and AIDS, according to an FDA compliance officer who helped to shut down production of the substance for several years.[41] Cancell is still marketed today under various names including Entelev, Cantron and Protocel.

Homeopaths also offer cancer patients highly diluted solutions of mistletoe, known as Iscador. The idea, based on homeopathy's Law of Similars and first suggested by Rudolf Steiner in the 1920s, is that because mistletoe grows on a tree like a cancer, it makes a suitable treatment for the disease. Or you could consult Hulda Regehr Clark, who believes that all types of cancer have the same cause, 'a certain parasite, for which I have found evidence in every cancer case' and which can be cured by her electronic 'zapper' in combination with the regular use of purgatives or 'organ cleanses'.[42]

Her clinic was closed in the US but she now operates in Mexico, in common with many other practitioners whose therapies are illegal elsewhere. Alternative therapists all over the world offer the 'Clark protocol' with its parasite zapping and 'cleanses'.

Usually the first thing that alternative therapists recommend is a radical change of diet. Barrie R Cassileth of the Memorial Sloan-Kettering Cancer Center in New York is one of the foremost scholars of complementary and alternative therapies for cancer. She describes how 'advocates of dietary cancer treatments typically extend mainstream assumptions about the protective effects of fruits, vegetables, fibre, and avoidance of excessive dietary fat in reducing cancer risk to the idea that foods or vitamins can cure cancer'.[43] Eating a mixed low-fat diet including plenty of fruits and vegetables may reduce the risk of getting cancer, but there is no evidence that it is possible, as one alternative recipe book has it, to *Eat to Beat Cancer* in that eating a diet containing a lot of fruit and vegetables will make a tumour go away.[44] [45]

But regardless of its worth or otherwise, the 'feelgood' diet is bound to be more attractive to the cancer sufferer than the decidedly 'feelbad' experience of orthodox cancer treatment. Twenty years ago Professor of Health Education William T Jarvis recognised how 'the "diet cure" appeals to the vulnerable cancer patient's need to re-establish control over himself and his body as much as possible.' But the risk is that patients 'may be tempted to abandon unpleasant radiation or chemotherapy in favour of the more emotionally satisfying, albeit ineffective, diet "therapy"'.[46] Despite the lack of evidence that dietary changes can affect the course of the disease, alternative nutritionists frequently suggest that people with cancer eat organic fruit and vegetables, that dairy products are substituted with foods derived from nuts or soya and that they should radically reduce consumption of meat, coffee, alcohol, salt and acidic foods such as citrus fruits.

One of the best-known alternative cancer dietary regimes is *Gerson Therapy*, devised by German émigré Dr Max Gerson in the 1920s and today promoted by his descendants. He began by claiming the diet could both cure migraine and TB, before going on to promote it as a treatment for cancer (TB in Europe having been rather effectively dealt with by 'organised medicine' – curious how there was no 'TB industry' to suppress a cure here). Gerson therapy is based on the erroneous belief that cancer is triggered by a weakened immune system, damaged by 'a variety of disease and cancer causing pollutants'. These toxins 'reach us through the air we breathe, the food we eat, the medicines we take and the water we drink', a suggestion that makes one wonder how there could be anybody in the developed world who has managed to avoid cancer.

In what is possibly the most foolish statement in all of alternative medicine, the Gerson Institute states that 'it is rare to find cancer, arthritis, or other degenerative diseases in cultures considered "primitive" by Western civilisation'.[47] Those cultures also have a low incidence of dementia, because like cancer, dementia is predominantly a disease of older age. The most recent UK statistics show that 74 per cent of people with cancer are over 60.[48] You have to live long enough to get it, and people living in developing countries often have too low a life expectancy to reach the age where they develop these conditions. (On reflection, this may not be the most stupid proposition in alternative medicine. Millionaire author and convicted fraudster Kevin Trudeau with his suggestion that 'animals in the wild don't get sick' tops that particular twit parade.)[49]

The Gerson Institute claims that 'with its whole-body approach to healing, the Gerson therapy naturally reactivates your body's magnificent ability to heal itself'. This advice can lead cancer patients to follow a diet that is demanding in the extreme. The regimen involves eating a diet high in fruit and vegetables: 'The aim is to flood the body with nutrients from almost 20 pounds of organically grown fruits and vegetables daily'. Let me repeat: *twenty pounds of fruit and vegetables daily*. A typical diet for someone on the full therapy regimen involves drinking thirteen glasses of fresh raw carrot, apple or green leaf juices each prepared hourly, three full vegetarian meals a day and constant snacking on more fresh fruit and veg. Followers also take an assortment of 'biological' medications, including potassium compound, Vitamin B12, pancreatic enzymes and injections of crude liver extract. They also have regular coffee enemas that are claimed to detoxify the tissues and the blood. In addition, they may be advised to use castor-oil enemas and oral and/or rectal hydrogen peroxide or rectal ozone treatment, live cell therapy, thyroid tablets, royal jelly capsules, linseed oil, clay packs, Laetrile (see below), and vaccines made from the influenza virus and staphylococcus aureus bacteria.

An intensive treatment at the Gerson clinic in Tijuana, Mexico costs approximately $5,500 per week and usually lasts around three weeks, after which the therapy routine is carried on at home for another eighteen months. The medication that accompanies the Gerson diet costs approximately $500 a month, and the recommended 'Norwalk' juicer (named after its inventor Norman Walker) costs $2,200.[50] The patient also has to buy organic coffee for the enemas, thyroid supplements, flax oil, special rye bread, a water distiller, an ozone machine and a set of 'non-toxic' household goods, such as cleaning products. It is estimated that up to sixteen hours a day will be taken up by shopping, preparing food and medication, administering enemas and cleaning the juicer. Juices, for example, may not be prepared in advance because they have to be drunk as fresh as possible.[51] Gerson therapy is perhaps the most demanding of all alternative cancer therapies and precludes having a job, let alone participating fully in family and social life.

Gerson therapy is backed by Prince Charles, who in 2004 told a London conference of health-care professionals 'I know of one patient who turned to Gerson therapy having been told she was suffering from terminal cancer and would not survive another course of chemotherapy. Happily, seven years later, she is alive and well. So it is vital that, rather than dismissing such experiences, we should further investigate the beneficial nature of these treatments'.[52] Evidently he was unaware of the numerous reviews and investigations of Gerson therapy over the years, none of which has shown it to be of any proven benefit in curing cancer. As long ago as 1947 two separate reports concluded that there was no evidence that the method worked. In 1959 an American analysis of Dr Gerson's description of fifty cases concluded that no benefit was proven, principally because of the lack of detail on the total number of patients – the suggestion being that Gerson had only written up the case histories of those who got better and had not reported his failures. Neither was there any detailed description of the grade and stage of the individuals' cancers or of their medical histories. Gerson's reports were therefore scientifically meaningless.[53] Gerson therapy is now illegal in the US when described as a cancer cure, which is why its Institute is in Mexico.

Not only is Gerson therapy ineffective but it can be positively harmful. During the 1980s there were reports of at least thirteen Gerson patients admitted to hospital with blood poisoning attributed to the liver injections. They were all infected with the organism campylobacter foetus, usually found in the intestinal tract of cattle and sheep. None was cancer-free and one died of his cancer within a week. Five were comatose due to low sodium levels, thought to be a result of the salt-free Gerson dietary regime.[54] Another report in 1983 tracked twenty-one patients who visited the Gerson clinic over a five year period (or until death). At the five year mark, only one was still alive – but not cancer-free – and the rest had died.[55] Other deaths have also been attributed to the coffee enemas, caused both by perforation of the bowel and disturbance of blood electrolyte levels.[56] Gerson's suggestion that coffee enemas stimulate the production of bile thereby enabling the body to detoxify, has been repeatedly disproven.

Other alternative cancer treatments are often incorporated into a Gerson-style dietary regime. Laetrile is made from the natural substance amygdalin which is found in the nuts and pips of many fruits, especially apricot kernels. It is also known as Vitamin B17, though it is not a vitamin. The substance was isolated in 1830 by two French chemists and first tried as an anti-cancer agent in Germany in 1892. No benefits were established and it was clear that the substance had unacceptably toxic effects – hardly surprising, since the active ingredient in Laetrile is cyanide.

After a number of court cases amygdalin has been banned in America and Europe because it has no proven curative effect and is so dangerous. Yet it is still available in clinics in Mexico and on the internet, often in a synthesised form as injections, tablets, capsules or enemas. It's also readily and legally available in its 'natural' state as apricot kernels, three kilograms of which cost £45 from the UK supplier Kernelpower. A pharmaceutical assessment of amygdalin/Laetrile published in the journal *Cancer Treatment Reports* found that doses of more than fifty kernels or 50mg of Laetrile can be fatal and that it is not safe to consume more than five apricot kernels a day.[57] But this finding is at odds with advice from internet sites such as anticancerinfo.uk which tells cancer patients 'to eat 1½ teaspoons of ground kernels (eg stirred into juice), or five or six whole (chewed) kernels, up to a maximum of six to ten times in a day . . . As more kernels are tolerated, it is suggested to aim for a maximum of 5 apricot kernels per waking hour up to a total of 50 kernels per day . . . it is important not to be half-hearted.'

Laboratory studies have shown that the cyanide in Laetrile can kill cancer cells but then you could also kill cancer cells in a laboratory with a flame thrower or a bottle of bleach. There have been no human studies that support its use.[58] In one 1982 trial one hundred and seventy-nine patients with advanced, untreatable cancer were administered a combination of Laetrile, vitamins and pancreatic enzymes. Measurable improvement was found in only one patient, 90 per cent had disease progression within three months and the median survival rate was less than five months.[59] There has also been considerable evidence of the effects of Laetrile-induced cyanide poisoning which include sickness, dizziness, liver damage, low blood pressure, fever, nerve damage, confusion, coma and death.[60]

Laetrile's most famous champion was the film star Steve McQueen, still known as the 'King of Cool'. McQueen was diagnosed in 1979 with mesolthelioma, an incurable lung cancer possibly caused by exposure to the asbestos in the protective suits he wore while car racing, as well as a sixty-a-day cigarette habit. He refused chemotherapy, but the prognosis was in any case poor. Secretly McQueen turned to dentist William Kelley, who claimed to have developed a cure for cancer, and went to Mexico where he was treated according to Kelley's prescription. As well as Laetrile, he was given pancreatic enzymes, an assortment of fifty daily vitamins and minerals, massages, prayer sessions, psychotherapy, coffee enemas and injections of a cell preparation made from sheep and cattle foetuses. After the US newspaper the *National Enquirer* revealed his story McQueen stated publicly that he was recovering and that

'Mexico is showing a new way of fighting cancer through non-specific metabolic therapies'. One of his doctors appeared at a press conference to claim that the Kelley treatment was achieving an improvement in 85 to 95 per cent of patents.[61]

Unfortunately Steve McQueen's cancer had spread. He died in 1980 from a heart attack following surgery to remove tumours from his abdomen and neck. Despite the death, Kelley continued to claim that Steve McQueen had been cured of cancer. He said this success had been intolerable for the 'Cancer Establishment' because McQueen 'was going to blow the lid off of the cancer racket' by publicising his cure. In an article entitled 'Cancer Cure Suppressed' he alleged that McQueen was murdered the night after his surgery by 'a government agent [who] came into his room posing as a physician on duty and injected McQueen with a blood clotting medication which was the cause of death.'[62] William Kelley died, discredited and despondent, in 2005.[63]

Despite its ultimate failure, Steve McQueen's pursuit of unorthodox treatment still has great cultural resonance, says Barron H Lerner, author of *When Illness Goes Public*. Within weeks of McQueen's death thousands of American cancer patients were seeking out alternative cancer treatment and many were 'clamouring for admittance' to Mexican hospitals. Even today ordinary cancer patients 'may see themselves as Steve McQueens, ready to marshal any possible forces against the dreaded disease,' writes Lerner.[64] This is another manifestation of the celebrity culture apparent across all media – just as we enthusiastically copy celebrity exercise and weight-loss programmes, we want to resemble the stars by using their glamorous-sounding alternative medicine practices, even when they fail.

As you might expect given the general assumption in alternative medicine that anything 'natural' is good, many alternative cancer treatments are based on herbal ingredients. Essiac is a herbal treatment given either by injection or as a tea made from four herbs: burdock, Indian rhubarb, sheep sorrel and slippery elm. It was created in the 1920s by a Canadian nurse, Rene Caisse, who claimed the formula was imparted to her by a native Ojibwa medicine man who said it could 'purify the body and place in balance the great spirit'. She named it after the backward spelling of her own surname. Caisse died in 1978 but the product is still widely available in North America and Europe.

Rene Caisse first proposed that the combination of these four herbs could strengthen the immune system, improve appetite, relieve pain and improve quality of life. Later she went on to claim that Essiac was able to attack a cancerous tumour directly, and ultimately to dissolve and expel it from the body. Caisse would not allow researchers properly to investigate Essiac, refusing to give away her 'secret formula'. But her unwillingness to cooperate with researchers did not stop her writing, years later, that 'I did not know then of an organized effort to keep a cancer cure from being discovered, especially by an independent researcher not affiliated with any organisation supported by private or public funds ... It would make these foundations look pretty silly if an obscure Canadian nurse discovered an effective treatment for cancer!'[65]

Clinical studies in the 1970s and 80s following up people who had taken Essiac showed no evidence of any change in expected disease progression.[66] The herbs contained in Essiac are not particularly toxic, but some people experience nausea, vomiting and diarrhoea and two of the herbs contain high concentrations of oxalic acid which could make it unsafe for those with kidney problems or arthritis.

Moving from the herbal to the more exotic end of the CAM spectrum, we find the popular anti-cancer remedy shark cartilage. It has been marketed as Carticin, Cartilade, Neovastat and Benefin. It is widely available, with 100 tablets selling in Holland & Barrett (motto: 'we're good for you') for £16.99. Since the recommended dose is sometimes as many as 12 capsules, three times a day, it's an expensive option. Shark cartilage is marketed as a food supplement and so is not regulated as a medicine. The shark cartilage supplement business is reported to be worth over $5 billion a year.[67]

Shark cartilage first came to prominence in the cancer field following the publication of the book *Sharks Don't Get Cancer* by Dr William Lane in 1992. This was followed by his imaginatively titled *Sharks Still Don't Get Cancer* in 1996. Lane's reasoning for why sharks didn't get cancer was based on fact that they have a high proportion of cartilage in their body. Because cartilage doesn't contain blood vessels Lane decided it could prevent the formation of a blood supply to solid cancer tumours. This idea has led to the wildly more ambitious claim that 'Taking Shark Cartilage gives the surrounding body the ability to "deny" the request for a blood supply. This is where Shark Cartilage can help in the fight against solid tumour cancer. It can help restrict new solid tumour growth by denying it a blood supply while the immune system and other cancer treatments fight the existing cancer'.[68]

The popularity of shark cartilage extract as an anti-cancer treatment is a textbook triumph of marketing and pseudoscience over reason, with tragic fallout for sharks and humans, according to Gary K Ostrander, research professor in Biology and Comparative Medicine at The Johns Hopkins University in America. He writes 'crude shark cartilage is marketed as a cancer cure on the premise that sharks don't get cancer. That's not true, and the fact that people believe it is an illustration of just how harmful the public's irrationality can be.' Ostrander and co-researchers have detailed more than forty published examples of tumours in sharks, some dating back to the 1800s.[69] To date, a total of eight uncontrolled trials and five controlled trials have produced no evidence that shark cartilage is an effective cure for cancer.[70] The most recent study, by the American Cancer Institute, found shark cartilage ineffective for lung cancer patients.[71]

Professor Ostrander thinks 'it is possible that highly purified components of cartilage, including from sharks, may hold some benefit for the treatment of human cancers. The key will be to isolate these compounds and design a way to deliver them to the site of the tumour. Lane and others ignore these existing barriers and suggest that consuming crude cartilage extracts . . . could be curative of all cancers, an approach for which there is no scientific basis'.[72] Furthermore, there is no recommended dose of shark cartilage and most over-the-counter products contain very little. Side effects can include hepatitis, jaundice and other liver problems, sickness, loss of appetite, diarrhoea, constipation and indigestion.

The sale of the shark cartilage supplement Benefin has been banned in the US since 2004. In a case initiated by the Food and Drugs Administration, two companies were charged with making false and unsubstantiated claims for shark cartilage as a cancer treatment. Andrew Lane – son of William Lane of *Sharks Don't Get Cancer* fame – was ordered to stop selling Benefin and to make restitution to anyone who had bought it since 1999. The illegal promotion of the product had continued despite an early FDA warning letter and a Federal Trade Commission cease-and-desist order. The judge described Lane and the other defendants as untrustworthy and issued a permanent injunction against their misrepresenting any product in the future.[73]

And what of poor old sharky in all of this? Studies suggest that the cartilage supplement market is seriously affecting the shark population and that there are grave implications for the marine environment. Sharks have a crucial role as the apex predator at the top of the food chain, where they maintain and determine the health of ecosystems. Marine expert Jean-Michel Cousteau says 'when the shark population declines sharply, the results can be dramatic, upsetting the balance of the oceans and producing un-intended consequences with effects that can reach around the globe'.[74] A 2007 study published in *Science* showed that if sharks are removed, the animals that they formerly fed on are likely to thrive, in turn wiping out species lower down the food chain.[75]

This assortment of largely discredited remedies represents only the tip, however, of a continent-sized iceberg of CAM treatments for cancer, as the American Cancer Society's advice on complementary and alternative cancer methods confirms. Though all the following are advertised in various countries in various forms and through a multitude of different outlets as cancer cures, there is no scientific evidence that any can cure or influence the course of any cancer: acupuncture, aloe vera, alsihum, antineoplaston therapy, applied kinesiology, aromatherapy, astragalus, aveloz, bee venom therapy, bioenergetic therapy, biofeedback, black cohosh, black walnut, bodywork, CanCell, cancer salves, cassava (tapioca), castor oil, cat's claw, cell therapy, centella (gotu kola), chaparral, chelation therapy, chiropractic, chlorella, colon therapy, comfrey, craniosacral therapy, crystals, cupping, curanderismo, cymatic therapy, dental amalgam removal, DHEA, Di Bella therapy, DMSO, electromagnetic therapy, enzyme therapy, Essiac tea, evening primrose oil, faith healing, fasting, feng shui, flaxseed, flower remedies, Fu Zhen therapy, germanium, Gerson therapy, ginkgo biloba, ginseng, goldenseal, grape cure, Greek cancer cure, guided imagery, HANSI, holistic medicine, homeopathy, Hoxsey herbal treatment, humour therapy, hydrogen peroxide, hydrotherapy, hyperbaric oxygen therapy, hypnosis, immuno-augmentative therapy, inosine pranobex, kampo, kombucha tea, krebiozen, labyrinth walking, Laetrile, larch, light therapy, lipoic acid, Livingston-Wheeler therapy, macrobiotic diet, magnetic therapy, maitake mushroom, meditation, metabolic therapy, moxibustion, mugwort, Native American healing, naturopathy, neuro-linguistic programming, Noni (morinda) plant products, ohashiatsu, oleander leaf, orthomolecular medicine, osteo-pathic manipulation, oxygen therapy, pau d'arco, polarity therapy, poly-MVA, potassium supplements, pregnenolone, psychic surgery, qigong, rabdosia rubescens, red clover, reflexology, reiki, Revici guided chemotherapy, Rosen method of bodywork, Rubenfeld synergy method, saw palmetto, sea cucumber, 714-x, shamanism, shark cartilage, Siberian ginseng, snakeroot, t'ai chi, Tui Na, urotherapy, visualisation, Vitae Elixxir, vitamin K, Watsu, wheatgrass products, and yoga.[76]

Not only are these remedies useless as a cure for cancer, some have serious side effects. There are also fears that some supplements, notably anti-oxidants such as beta-carotene, may interfere with the action of orthodox treatments, particularly chemo and radiotherapy, making them either less effective or dangerously toxic.[77]

In the UK the regulatory framework may be theoretically robust but it is under-policed to the extent that there is now inadequate protection for the cancer sufferer. The Cancer Act of 1939 prohibits anyone other than registered medical practitioners from claiming they can cure or treat cancer. This legislation also covers the adver-tising of quack cancer cures, but it is up to local Trading Standards Authorities or the Medicines and Healthcare products Regulatory Agency to initiate any prosecution, and conviction typically results in a small fine or three months imprisonment. Prosecutions are few. And whilst the UK National Cancer Research Institute has recently established a Complementary Therapies Clinical Studies Development Group, this body was set up in order 'to promote research into complementary and alternative medicines in cancer care' and not to police or provide guidance on what is already available. Jonathan Waxman, Professor of Oncology at Imperial College London, believes it is time for a revision of the legislation of complementary remedies directed at cancer patients, whose supporting claims 'may be overtly or malignly incorrect'. He suggests they should not be categorised as food supplements and should be properly tested for effectiveness. 'Reclassify these agents as drugs – for this is after all how they are marketed – and protect our patients from vile and cynical exploitation whose intellectual basis, at best, might be viewed as delusional.'[78]

How cancer quackery 'works'

It's easy to be persuaded by tales of miraculous cancer cures attributed to alternative treatment methods, but things are very rarely as they seem. Here's how cancer patients can be, and often are, misled into thinking that alternative cancer treatment works:

First tell the patient they have cancer – then say you have cured it. This is easy when the therapist is using a transparently dubious diagnostic technique like Robyn Welch who intuitively tracks your ailments via the telephone or Hulda Regehr Clark with her 'synchrometer'. Patients can leave happy, believing they are miraculously 'cured' of a cancer they never had in the first place. This practice is what saw Clark being run out of the US after she falsely diagnosed an Indiana Department of Health undercover agent, Amy Huffman, with AIDS. She told her 'I can test you here. It's a one minute test and it's all electronic'. She later told Huffman she was 'full of the virus. And we will cure it in three minutes'.[79] Sometimes the patient diagnoses, as in this example from a testimonial in support of the 'cancer salve' Cansema: 'I had two lesions that I was concerned about for about a year, one on my arm and one on the top of my head . . . I did not have them checked by a doctor since I was trying other health alternatives but I am sure they were some form of skin cancer'.[80]

A cancer is cured or put into remission by orthodox medicine, but the credit is given to CAM. One woman's 'successful journey', published on the curezone website (motto: 'educating instead of medicating') is a classic example of this kind of selective attribution. Nan McSweeny admits to having surgery for her early stage breast cancer, but is determined to attribute her continuing remission to a time-consuming alternative medicine regime. 'I went in for lumpectomy, and that is where I stopped. I did research, meditated (I am a reikimaster and meditate every day) and prayed . . . I did intensive self-development, energy healing, bodywork and nutrition . . . I closed my business and have become passionately involved with some nutritional products . . . I also tried macrobiotics, did a lot of brown-ricing. I did wheat grass enemas, and heavy metal detox, I am still taking 2–3 oz of wheat grass every day orally, and do enemas once a month'.[81] But early stage breast cancer of the kind described by Ms McSweeny has a good prognosis if treated promptly with surgery, even without the use of adjuvant treatments such as chemotherapy, radiotherapy and other drug treatments that would have further improved her chances.

Penny Brohn, one of the founders of the Bristol Cancer Help Centre (now called the Penny Brohn Cancer Help Centre) recorded her experience of recurrent breast cancer and her use of complementary therapies in her book *Gentle Giants*, describing how, despite her refusal of radiotherapy, her tumour began to shrink.[82] Writing in 1993, oncologist Dr Robert Buckman and writer Karl Sabbagh noted 'although there are not claims of miracles [in the book], there are many references to how surprising the remission was, and the reader could be left with the very strong impression that the reversal of the prognosis . . . must in some way be related to the many complementary treatments which take up so much of the book. There are only six sentences in the whole book that give what we believe to be a more accurate real answer.'

When Penny Brohn's cancer recurred it was found to be responsive to the orthodox hormone therapy tamoxifen, which she took for seven years. 'It is in no way surprising that the recurrence shrank when she continued to take the most active conventional anti-cancer drug in that situation,' say Buckman and Sabbagh. Whilst they never suggest any attempt at active deception in Brohn's book, they point out that it 'reads like a story of the triumph of complementary medicine, a story in which the real hero, tamoxifen, does not get its deserved praise'.[83] Penny Brohn died in 1999 after living with cancer for nearly twenty years.

Alternative practitioners can get themselves in considerable tangles when trying to discount the positive effects of orthodox medicine. Here is one strangulated explanation of why patients receiving chemotherapy might do better than those who haven't received it: 'A person who took chemo and suffered the usual debilitating side effects has extremely high motivation to change his or her diet and lifestyle. A person who took a side-effect-free alternative cancer treatment may not feel as strongly motivated and may not make the changes needed to prevent reoccurrence. Therefore, how long a person remains tumour free after taking a cancer treatment is *not* an accurate measure of the effectiveness of the treatment. It is more likely a measure of how well that person changed their lifestyle'.[84] It wasn't the chemotherapy that cured you – it was the diet, stupid.

The cancer is said to be cured but is actually progressing. Sadly, this often seems to be the case with celebrities who reject mainstream medicine and whose use of alternative cancer therapies is publicised. Of course this can happen to anyone, but it is the famous whose often public 'battle against cancer' attracts the airtime, acting as an advertisement for something that eventually fails.

UK television presenter Caron Keating refused tamoxifen after undergoing initial orthodox treatment and instead turned to Brandon Bays, whose 'Journey therapy' promises to bring about healing by uncovering 'specific cell memories, resolve them completely and clear them out. Your body and being then go about the process of healing quite naturally, automatically and you are left soaring in a boundless joy, peace and wholeness that is your own essence.'[85] [86] You may remember that Bays claims she was herself diagnosed with a uterine tumour 'the size of a basketball' in 1992. Bays says that when her surgeon advised immediate removal she decided to use natural methods to reduce the tumour's size, attesting that she took no drugs, underwent no surgery and instead used 'noninvasive treatments including vitamins, a radical change in diet, massage and various other emotional and physical therapies. Miraculously, within a six and a half weeks, the tumour disappeared'.[87]

Brandon Bays has said she is inspired by the work of Deepak

Chopra, and there are certainly echoes of his pricing structures and recruitment techniques in her project. 'Journey Intensive' two-day seminars are held all over the world, for which she charges each of the reportedly five hundred or more attendees per event £245.[88] After one of these 'you become a Journey Grad which opens you to a broad range of benefits and support and qualifies you to attend the advanced Journey programs' such as the two-day Manifest Abundance Retreat, which costs £670. There are Journey-trained practitioners all over the world with more than one hundred in the UK. Caron Keating did not make her cancer publicly known, but moved to Byron Bay in Australia, a place her mother Gloria Hunniford later described as 'the shopping mall of the world when it came to alternative therapies', and, with Brandon Bays, made the DVD *Instant Calm*, a beginners guide to relaxation, yoga and meditation.[89] Tragically, the cancer recurred in her other breast but she declined further surgery, choosing instead to pursue alternative treatments in Australia and Switzerland. In the end the cancer spread to her bones and could not be stopped by any means. She died in Britain in 2004 leaving a husband and two young sons.[90]

When the British motorcycle champion Barry Sheene fell ill with stomach and oesophageal cancer in 2002 he talked openly about his refusal of orthodox treatment and his belief in alternative medicine. He went on a diet that consisted entirely of beetroot, Chinese cabbage, carrot and radish juice, saying he was 'fighting cancer the natural way'. Sheene told a reporter from an Australian newspaper that he had refused surgery and chemotherapy because 'anybody I've ever known who's had it has been, basically, completely destroyed. I can't let someone put an IV drip in my arm and inject me with poison. I've seen friends of mine after chemo. They look like dead people who can still talk'.[91] Barry Sheene, in the lingua franca of modern cancer discourse, 'lost his battle with cancer' only six months after diagnosis, at the age of fifty-two. Whilst it is impossible to predict how long Sheene might have survived had he been treated with orthodox medicine, current statistics show that with treatment the overall five-year survival rate for stomach cancer is around 20 per cent. Early diagnosis of stage 1 stomach cancer results in a 70 per cent survival rate at five years.[92]

The Dutch actor and comedian Sylvia Millecam was diagnosed with breast cancer in 2000. Even though she had a small, early-stage tumour she refused the recommended surgery and instead saw a psychic medium who told her that she did not have cancer and advised against chemotherapy. She went on to seek various cures, including something called 'cell specific cancer treatment' at a Swiss clinic, salt therapy, electro acupuncture, homeopathic and psychic healing. Once again she was told that a Vegatest showed that she did not have cancer but a bacterial infection. Instead of prescribing conventional drugs for pain relief her doctor used 'magnetic field apparatus' in order to try to identify a blood disorder. It is believed Sylvia Millecam consulted at least twenty-eight different alternative practitioners and institutions in the year between diagnosis and when she died in 2001.

Following Millecam's death the Amsterdam Medical Disciplinary Tribunal struck off two of her doctors, and suspended the third for a year. They were judged to have ignored existing standards for treating breast cancer, to have used unsatisfactory methods and to have withheld information. Whilst the doctors were not held responsible for her death, the tribunal found the first two had denied Ms Millecam a reasonable chance of recovery by failing to refer her to a breast cancer specialist. The third was judged to have caused Millecam unnecessary suffering by withholding conventional palliative care when her condition deteriorated. The tribunal judge felt it necessary to repeat that a doctor who explores the terrain of alternative medicine nevertheless carries a doctor's responsibilities.[93] [94]

Cancer is unpredictable. The general public is not well informed about cancer. There is a widespread assumption that a cancer diagnosis is a death sentence and that every cancer is equally dangerous. In fact there are as many as two hundred different types of cancer. The typical course of each is different and the actual course will vary from person to person. Even without orthodox treatment it can take years for the disease to progress and there may be remissions lasting many months or even years. In forms with a very poor prognosis, such as pancreatic cancer, there will always be one or two people in every hundred who defy the odds, live for longer than expected or who have a spontaneous remission. It is in these cases we are most often encouraged to attribute survival to the patient's attitude, lifestyle or use of alternative treatments, even though there is no evidence that any of these can improve survival rates.

Desperate measures

We are all vulnerable to those who seemingly offer a way out of what seems a hopeless situation. In 2005 Frank Bagyan, a Canadian middle-aged father of three grown-up daughters, began to suffer blinding headaches. He was diagnosed with an inoperable brain tumour and told he would probably not live beyond six months. Distraught at the news, he and his wife Teresa were relieved when they came across the professional-sounding Canadian Cancer Research Group. Former computer consultant Bill O'Neill, who despite holding no medical qualifications is known as 'Dr Hope', runs the CCRG. O'Neill, according to Teresa Bagyan, 'looked Frank in the eye and told him he had never lost a patient yet'. She describes how O'Neill said that her husband had come in plenty of time and that the treatment would dissolve the brain tumour.

Frank Bagyan paid $5000 in advance, had blood samples taken and was given a collection of nutritional supplements including vitamins and flaxseed oil. Bill O'Neill calls this method Aminomics© which the CCRG says 'involves the application of proprietary diagnostics in the form of blood analysis that drives the manufacture of bio-individualized therapeutic compounds'.[95] Its appeal is obvious. O'Neill also says the CCRG is 'at the forefront of cancer research' and that it sees more than two thousand patients a year, each paying up to $10,000 for treatment. The fees are justified because 'the centre has achieved one year cancer survival rates that hugely exceed the national norm . . . we are literally melting off tumours in many cases'. There is no published evidence to support O'Neill's claims, perhaps, as he puts it, because the CCRG's services are 'beyond the formularies of socialised medicine'.[96]

A few weeks later the couple were delighted when Bill O'Neill told them that Frank's latest MRI scan showed the treatment was working and the tumour was starting to break up. Just one week after this consultation, Frank collapsed and died of his brain tumour. Painful as it was, Teresa Bagyan related what happened in a Canadian TV documentary exposé of Bill O'Neill and the CCRG.[97] She accused

O'Neill of 'offering us false hope, he took advantage of our desperation. You have to walk a mile in a desperate person's shoes and then you know how people can be so stupid.'

After the programme O'Neill and the Canadian Cancer Research Group released this statement: 'Our view of many of the matters publicly raised by a few family members is that these individuals who are grieving have been exploited by the detractors of complementary medicine for the detractors' own benefit (financial and ego).' Bill O'Neill still promotes Aminomics© from his Ontario clinic and his website. He says he is suing the television company who exposed him.[98]

The lengths to which the sick will go were again made horribly clear when in 2003 the BBC Wales *Week In Week Out* programme investigated the activities of two alternative remedy suppliers, Jim Wright and Roy Mackinnon. Mackinnon was reported as advising cancer patient Eileen Murdoch that she needed to have all her teeth removed by a Swansea-based 'holistic dentist', John Anderson. The treatment cost her £2,500 and involved being away from home for ten days. Her husband said 'it was a very miserable time. I didn't for a minute believe she was going to be cured using the method, but I just thought "we'll go for it" because you can't spend the last months of their life fighting with your partner, can you? Bearing in mind that was about five weeks before she died of terminal cancer she was in a bad way. I was going along with it but I was in a daze. I remember her sitting in the B&B with a very swollen mouth, dribbling, with terrible headaches.'[99]

Wright and Mackinnon were subsequently investigated and prosecuted both under the Cancer Act and for illegally selling unlicensed medicines by Trading Standards and the Medicines and Healthcare products Regulatory Agency. Roy Mackinnon was offering what he called the 'Hulda Clark toxin/pathogen removal process', but his prosecutors could produce no evidence that he had actually said he could cure cancer, so the case collapsed. He later told the *Institute of Complementary Medicine Journal* that the case against him had been backed by 'the pharmaceutical multinationals' because 'no pharmaceutical company that has invested hundreds of millions of dollars in a supposed cancer cure wants the public to get to hear of a natural approach', but offered no proof for such a conspiracy.[100]

Jim Wright sold remedies – stored in his shed – via his websites Goodbye2Cancer and Goodbye2Pain, telling prospective customers they could 'fight all types of cancer'. Wright was found guilty of six counts of illegally selling medicines including Laetrile and a product called Quikheal Green. He was sentenced to 120 hours of community service and to pay £1,000 in costs. Following the sentence, Mick Deats, Head of Enforcement and Intelligence at the MHRA, said 'Mr Wright was blatantly deceiving the public and putting their health at risk by selling these unlicensed medicines. He sold medicinal products to vulnerable members of the public – all for his own profit and gain. These medicines have absolutely no guarantee of safety and no evidence to show that they alleviate the conditions they claimed to treat'.[101]

There is a view that one should not criticise anything that provides hope for cancer patients, but it is possible to be hopeful and positive while still being realistic. American doctor Jerome Groopman gives perfect expression of this when he writes about the difference between false and true hope: 'false hope does not recognise the risks and dangers that true hope does. False hope can lead to intemperate choices and flawed decision making. True hope takes into account the real threats that exist and seeks to navigate the best path around them'.[102]

Australian Lesley Bramston died two days after she was injected with a caesium chloride treatment – sometimes called ozone therapy – recommended by alternative practitioner Paul Rana and his NuEra Care Clinic – 'Bridging The Gap Between Ancient Wisdom & 21st Century Health Care' – in Melbourne. Rana says that with cancer cases 'nutrition should be the first point (sic) of call. Only after giving nutrition and complimentary (sic) therapies a powerful opportunity to work, should surgery be considered'. He 'does not recommend chemotherapy, because in years' (sic) involvement in this field, he has not seen any benefit in it, except palitive (sic).'[103] Instead, 'The Rana System' offered vitamin and mineral supplements, Laetrile, caesium or high PH therapy, parasite/energy zapping devices, Zen Chi Massages, magnetic pulsers, coffee enemas, ozone therapy, 'blood type' diets, live blood analysis and thermal imaging.

Ms Bramston, a 44-year-old mother of two, was suffering from incurable kidney cancer. Her husband Paul McNamara says he paid Rana A$44,000 for vitamin and herbal treatments for his wife, who was told that additional ozone therapy 'would have her dancing in a couple of months.' Rana referred her to Dr Hellfried Sartori and his clinic in Australia's Northern Territory. Bramston's sister described what happened next. The therapist 'took this solution out of an oxygen bottle and injected it into an IV in [Lesley's] arm. Her left side went rigid like she had had a stroke and her right side was in spasm. Then she slowly slipped into unconsciousness . . . for the next two days she dropped in and out of consciousness, making sounds and murmurs. On the Wednesday she wasn't getting any better. They tried to lay her down and she started making these noises and then she stopped breathing. The nurse tried to give her CPR. After a while they called an ambulance. The officers tried to revive her, but she was gone'.

After her death her husband Paul told the *Weekend Australian*: 'I feel like a fool for letting my wife go up there . . . I let her down. But he (Rana) was behind it all. He is playing with people's lives and using their money.' After Lesley Bramston died her husband received a further bill for A$25,000 which he refused to pay.[104] In 2007 the Australian Federal Court imposed a permanent restraint on Paul Rana and his sons Micheal (*sic*) and Lee from peddling fake cancer cures and ordered them to pay court costs. The court found they had 'engaged in misleading or deceptive conduct' of 'the most reprehensible kind, revealing a cynical and heartless exploitation of cancer victims and their families when at their most vulnerable'.[105] In 2008 Paul Rana was sentenced to six months in prison. It turned out that Hellfried Sartori – known as 'Dr Ozone' – had been jailed in New York State in May 1992 and Washington in July 1998 for illegally administering his so-called 'ozone treatments'. In 2006 he was charged by Thai authorities with fraud and practising medicine without a licence in the northern city of Chiang Mai following the deaths of several of his foreign patients.[106] [107]

The German husband of Michaela Jakubczyk-Eckert, who died of breast cancer in 2005, is determined to publicise the dangers of rejecting orthodox medicine in favour of an alternative approach. Fearful of a cancer diagnosis, Michaela delayed seeking treatment for a breast lump but once on chemotherapy appeared to be responding well. In 2002 she decided however to consult alternative practitioner Dr Ryke Geerd Hamer, at that time based in Spain, who avows that 'all diseases ultimately have a psychic origin' and that it is 'only modern medicine that has turned our animated beings into a bag full of chemical formulas'.[108] Hamer recently spent eighteen months in a French prison after being convicted of fraud and of practising medicine without a licence.[109] According to her husband Gilbert Jakubczyk, Hamer made Michaela believe 'that conflict is the cause of the illness, the cancer being the healing phase' and persuaded her to stop chemotherapy. Gilbert was so opposed to this that the couple became estranged. Following Michaela's death in 2005, Gilbert posted a harrowing obituary on the internet as a warning to others, together with shocking photographs of Michaela's advancing cancer, which by the end of her life covered most of her upper body. He described how Michaela became so ill that a visiting doctor ordered her to be taken to a hospice immediately. 'Never before the personnel of the hospice had seen such misery, a human, who – alive – rots and is only mere skin and bones . . . despite strongest medications she cried of pain, she cried down the whole house . . . She lived on for only four days.'[110]

These tragic accounts show how terrified people can be persuaded to believe and do almost anything, all in the name of taking responsibility for their health and seeking a miracle. Not only are fortunes extracted from them, but cancer patients are also deprived of comfort and palliative care at the end of their lives. They and their families have to suffer the misery and humiliation that accompanies the dashing of false hope and the knowledge that one has been 'had', as expressed so movingly by Teresa Bagyan. And in one last, quiet crime against people with cancer, alternative medicine strands patients and their families in a state of denial and prevents them from adjusting to the realities of the situation. This, as William Jarvis has written, removes the opportunity for some satisfaction in the final days, depriving family and friends of the chance to accept and say goodbye.[111]

Chapter 8

Fads, Fashions and Faking it

Just as alternative medicine offers a complete wardrobe of therapies and remedies, it also boasts a set of matching accessories in
the form of its own illnesses and conditions. For as well as claiming
to treat conditions from back pain to cancer, alternative medicine
diagnoses a panoply of ailments largely unrecognised by orthodox
medicine. At the greatest extreme a whole discipline may be built
on a nonexistent phenomenon like the founding myth of chiropractic: the spinal subluxation.

The advantage to the alternative medicine practitioner is obvious:
a fabricated illness is much easier to diagnose and treat than a real
one. In due course you can tell the patient that they are cured and
thereby add to your reputation, or better still say they need further
treatment or perpetual 'wellness care' and add to your bank balance.
Either way, the result is a healthy flow of repeat custom. The benefit
to the patient is less apparent, but the naming and consequent
endorsement of human experience's more difficult manifestations
can be hugely gratifying, especially if you feel these are neither
recognised nor fully appreciated by mainstream medicine. Once
acknowledged in this way, the invalid becomes, in their own mind
and those of others, validated.

Many of the new conditions that have emerged in recent decades
cannot be defined as diseases, because they have no clinically measurable signs or symptoms. Instead they are illnesses or syndromes,
real only in a subjective sense to the sufferer. But there are plenty
of fad diagnoses to choose from – ones caused by the effects of
stress, electro magnetic radiation and unspecified 'toxins' to chronic
conditions such as fibromyalgia or multiple chemical sensitivity. What's

more, genuine medical conditions are often appropriated and diagnosed by alternative practitioners despite the patient exhibiting no signs or symptoms.

These fad conditions can be split broadly into two. The first kind are those unknown to mainstream medicine but often diagnosed by alternative practitioners, for example candidiasis hypersensitivity, leaky gut syndrome or generalised enzyme deficiency. Diagnoses of such conditions are frequently made by therapists using quack devices such as EAV or Vegatest machines and by other discredited methods like hair analysis. Then there are the real conditions that are extensively over-diagnosed by altmed therapists, like food intolerances and parasite infestation. The frequent lack of objective confirmation of these conditions is no hindrance because, as one specialist who claims a 91 per cent success rate in treating chronic fatigue syndrome cannily puts it, 'we don't treat the blood test, we treat the patient'.[1] Indeed the absence of symptoms may even be presented as proof of a problem, as one 'parasite cleansing' sales pitch shows: 'the immediate question that comes to mind when people are informed of this situation is: how can a parasite possibly live in my body and I don't even know it is there? The answer to this is simple. The purpose of a parasite is to not make itself known. A smart parasite lives without being detected because if it is detected, of course, something is going to be done to eradicate it. If you think parasites are stupid, think again'.[2] To the determined salesperson, absence of evidence is evidence of presence.

Imaginary friends

Morgellons disease was first described in America in 2002 by Mary Leitao after her two-year-old son Drew started complaining of bugs in his skin. Her doctor diagnosed eczema, but Mary Leitao believed her son was infected by something very different and much more threatening. She discovered the term Morgellons in a letter written in 1674 by the British writer and philosopher Sir Thomas Browne that described something similar. 'Hairs which have most amused me have not been in the Face or the Head, but on the Back, and not in Men but in Children, as I long ago observed in that Endemial Distemper of little Children in the Languedock, called the Morgellons, wherein they critically break out with harsh Hairs on their Backs, which takes off the Unquiet Symptomes of the Disease, and delivers them from Coughs and Convulsions.'[3]

The fact that such a condition had not been identified by orthodox medicine in the intervening three hundred years did not deter Ms Leitao from setting up the Morgellons Research Foundation, dedicated to finding the cause of this 'emerging infectious disease. We refer to this infectious disease as "Morgellons Disease", due to the need for a consistent label when contacting politicians and health departments'.[4] Since the establishment of the MRF there has been widespread reporting of examples of this 'frightening and painful infectious disease', most of which occur in California, Florida and Texas though there have also been reports of it in Canada, the UK, Australia and the Netherlands. As well as experiencing what they describe as 'crawling sensations' within and on the skin surface, sufferers see fibres and hairs protruding from their skin – they may gather fuzzballs, grains of a sand-like substance or black granules, as well as developing skin lesions and sores. But these are just some of the 'Six Signs or Symptoms' described by the MRF, who say patients are also likely to experience fatigue, cognitive difficulties or behavioural effects such as Attention Deficit Hyperactivity Disorder (ADHD) and Obsessive Compulsive Disorder. People with Morgellons also report joint pains, irritable bowel syndrome and changes in skin pigmentation. Some will have already been told they have other conditions frequently diagnosed by alternative practitioners, such as Lyme disease, fibromyalgia or chronic fatigue syndrome. The MRF has set up a self-reporting registration system in order to keep track of Morgellons around the world. By 2008 nearly thirteen thousand households had registered.[5]

In a *Times* interview in 2006 Dr Norman Levine, Professor of Dermatology at the University of Arizona, described seeing one hundred patients who believed they had Morgellons who had responded well to treatment which included the use of the antipsychotic drug Pimozide, used to treat mental illness with symptoms such as hearing, seeing or sensing things that are not there, mistaken beliefs and problems dealing with other people.[6] [7] Dr Levine's suggestion is that Morgellons patients suffer from the psychiatric condition 'delusional parasitosis', a disorder in which a person is absolutely but falsely convinced that there is some alien and parasitic entity living on or inside them. He described how 'he has studied the fibres patients bring in by the bag-load and they are textile in nature'. Some clinicians have suggested that the sores of Morgellons sufferers may be self-inflicted, caused by scratching an area the patient believes to be infected.[8] Other dermatologists and psychiatrists have endorsed Dr Levine's view.

In response to several years of pressure from the Morgellons patient lobby, the US Centers for Disease Control and Prevention (CDC) has recently decided to investigate the condition in order to determine whether or not it exists. CDC spokesman Dan Rutz said 'not a day passes when I don't talk to somebody who claims to have this. In the absence of any objective review, people have jumped to conclusions and found each other on the internet and formed their own belief structure. We really need to debunk this if there isn't anything to it or identify if there is indeed a new, unrecognised disease that needs attention'.[9]

Though dermatologists and others may anticipate a thorough debunking of Morgellons by the CDC, this is unlikely to change the minds of those who believe they have contracted it and who are unfazed when those they characterise as the 'medical establishment' reach such a conclusion. They are already likely to perceive their doctor's failure to diagnose them with Morgellons as consistent with the general failure of orthodox medicine to address their needs. Sufferers are especially angered at the suggestion that they could be released from their affliction by psychiatric treatment. As the National Unidentified Skin Parasite Association (NUSPA) puts it 'the medical community finds it much easier to label people with USPI (Unidentified Skin Parasite Infection) as delusional or offer a diagnosis for a disease which has nothing to do with what is really going on, than examine patients properly and look for a correct diagnosis and/or effective treatment'.[10]

Instead sufferers' beliefs are reinforced by alternative practitioners who encourage the use of treatments including oral doses of ionic silver and copper, eating pomegranate and using aromatherapy oils. The oils can be bought direct from an internet store which claims 'they penetrate every cell in your body within twenty minutes to kill all bacteria, fungus and viruses'. They also recommend candida cleanses, regular washing with borax and air purifiers 'to remove [the parasites] from the air as they leave your body'. But be warned. 'This is not an overnight kill. It could take two years depending on the amount of your infestation.'[11]

Scattergun syndrome

Prepare to diagnose yourself with Wilson's Syndrome, aka Wilson's Temperature Syndrome or even WTS, first described by American doctor E Denis Wilson in the 1980s. It's caused by what he calls 'functional hypothyroidism', a form of low thyroid function. Is there anyone who hasn't experienced several of the following, especially those who are female, middle-aged and who also happen to be the biggest consumers of alternative medicine? 'Fatigue, headaches, migraines, pre-menstrual syndrome, weight gain, depression, irritability, fluid retention, anxiety, panic attacks, hair loss, poor memory, poor concentration, low sex drive, unhealthy nails, dry skin and hair, cold intolerance/heat intolerance, low motivation, low ambition, insomnia, allergies, acne, carpal tunnel syndrome, asthma, abnormal swallowing sensations, constipation, irritable bowel syndrome, muscle and joint aches, hives, asthma, slow healing, sweating abnormalities, Raynaud's phenomenon, itchiness, irregular periods, easy bruising, ringing of the ears, flushing, bad breath, dry eyes, blurred vision.' The clincher is a body temperature that averages below 37°C (98.6°F). Since 'normal' can be anywhere between 36.5°C and 37.2°C (97.7°F and 99°F) and depends on time of day, activity levels and other variables, a diagnosis of Wilson's Syndrome can probably be made for just about everyone at some time or other. But the low thyroid function that Wilson says is the underlying cause of the condition cannot be identified by any known laboratory test. And after a thorough research review the American Thyroid Association 'found no scientific evidence supporting the existence of "Wilson's syndrome".'[12]

Following the death of a 50-year-old woman from a heart attack after he had prescribed her excessive amounts of thyroid hormone medication, a malpractice suit was filed against Wilson. His licence to practise medicine was suspended in 1992 by the Florida Board of Medicine who also fined him $10,000, but Wilson is still active on the internet, promoting his syndrome with case studies and testimonials, as well as selling a variety of herbal and nutritional supplements, such as 'Thyrocare'.[13] In the UK hypothyroidism is championed by 'The Thyroid Patient Advocacy Group' and by a private practitioner, Dr Barry Durrant-Peatfield, who conducts clinics countrywide and is the author of *The Great Thyroid Scandal and How to Survive It*.[14] [15]

In the past Dr Durrant-Peatfield offered treatments that were actively opposed by endocrinologists, including the unlicensed product ArmourThyroid which is made from the thyroid gland of pigs. But in 2001 he was suspended from the medical register by the General Medical Council so cannot therefore provide his patients with prescription-only medicines.[16] These can still be prescribed by another UK hypothyroid therapist, Dr Gordon Skinner, but following an investigation by the General Medical Council his registration has been made conditional and subject to a number of strictures, including the requirement that he should provide endocrine treatment only to people who have been referred by their GP, and that he should record and clinically justify in writing any prescribing outside recommended guidelines.[17] According to their supporters these two doctors have been subject to a witch hunt, and they have responded by mounting a campaign to abolish the GMC.[18]

Sick note

Despite the regular statistics produced by authorities like the World Health Organisation that show that we are healthier than ever before, the Western educated middle classes have developed such a heightened degree of bodily awareness that they are likely to interpret any unusual physical sensation as a sign of organic disease. This hypervigilance is a modern phenomenon. A poll of American households in the 1920s reported 82 episodes of illness from all causes per 100 of the population. A similar survey conducted in the 1980s reported 212 illnesses per 100.[19] This 158 per cent rise in illness perception not only defies the advances in health care during the twentieth century but shows how what could be described as the sickness threshold has plummeted.

Governments across the rich world struggle to reduce the numbers of those unable to work because of pain and fatigue-related conditions to which modern medicine has no answer, or at least no answer that the sufferer is prepared to accept. Modern consumerism, family breakdown, increased social isolation and preoccupation with ourselves have together led us into a new and sophisticated kind of hypochondria, blighting our lives with a collection of debilitating disorders. As one doctor puts it, 'headaches, heartaches, backaches, aching feet, fatigue, anxiety and those vague burning pains in your legs at night – these are the nemeses of real doctors. Many people have these symptoms, but the cruel truth it that there is no reliable cure for any of them'.[20] Surveys show that 80-95 per cent of us experience at least one such symptom every two to four weeks and the typical adult has at least one symptom every four to six days.[21] What has changed in recent years is that we are no longer reassured when we are told by an orthodox doctor that we are perfectly healthy or she can find nothing wrong. Today we shop around until we find someone who will provide a diagnosis and put a name to these distressing sensations and bodily events. Complementary and alternative medicine is more than willing to do just that.

The current epidemic of fad illnesses reveals much about the strategies of the alternative practitioner as well as the place of illness and the ill person in the modern world. Fad conditions are highly contagious, increasingly transmitted not by old-fashioned means like physical contact, but through the medium of the internet. Many are described extensively on American websites, only to become rampant in Europe a few years later, most often in the UK where the common language facilitates infection. Nor does it take long for a fad disease to take hold. Medical historian Professor Edward Shorter of the University of Toronto describes how they 'appear first among educated people simply because they are more plugged into medical media. These middle and upper class people are the first to begin monitoring themselves or their children for evidence of peanut-butter allergies or excessive tiredness. It is from these relatively small groups that the symptoms radiate'.[22]

In his fascinating book *From Paralysis to Fatigue*, Professor Shorter describes a 200-year history of illnesses that have been constructed, by both doctors and their (predominantly female) patients, to reflect the medical and social interests of the day. The medical profession of the nineteenth century already had a considerable stake in women's illnesses and in perpetuating the myth of female frailty, as Barbara Ehrenreich and Deirdre English point out in their pamphlet *Complaints and disorders: the sexual politics of sickness*. The myth had two purposes for male doctors: 'it helped them to disqualify women as healers, and, of course, it made women highly qualified as patients . . . a few dozen well-heeled lady customers were all that a doctor needed for a successful urban practice'.[23]

Edward Shorter describes how medicine and culture interact to generate an ever-changing assortment of symptoms. After the Edinburgh physiologist Robert Whytt demonstrated the reflex action of the spinal cord in the 1750s, patients became oriented

Mrs Bradbury's establishment for the reception of
ladies nervously affected, 1850

'away from their bowels and abdomens, where humankind had been fixated for about 500 years, to the central nervous system. Rather than talking about dysphorias in terms of the spleen, patients would increasingly start talking about nerves,' says Shorter.[24] By the 1800s many patients were being diagnosed with 'spinal irritation' that manifested itself in the form of paralysis, blindness and sleep-walking. It was diagnosed using a technique that was very similar to the 'tender point' method employed in the twentieth century for the diagnosis of fibromyalgia.

Later other novel complaints like 'railway spine' began to be diagnosed in parallel with technological advances. As science learned even more about the central nervous system diagnoses of 'American nervousness' or 'neurasthenia' appeared, only to be replaced in the late twentieth century by similar symptoms – notably fatigue and muscle pain – being ascribed to immune disorders, reflecting the increasing awareness of the immune system.[25] Shorter suggests that these interests acted as templates on which symptoms could be shaped by both doctors and their patients. This is arguably still the case, with modern medical specialists tending to diagnose and label the same group of symptoms according to their own area of interest, so a neurologist will diagnose a neurological disorder, a gastroenterologist diagnosing irritable bowel disorder, an allergy specialist food intolerance and so on.[26]

Today there is no greater perceived put-down than 'it's all in your mind'. A recent survey by the UK National Centre for Social Research confirmed much greater prejudice against people with a mental illness than those who have a physical one.[27] So in a world where mental problems are so cruelly stigmatised, a diagnosis of organic disease has become vital to our self-esteem. As feminist critic Elaine Showalter writes in *Hystories*, 'the culture forces people to deny the psychological, circumstantial, or emotional sources of their symptoms and to insist that they must be biological and beyond their control in order to view themselves as legitimately ill and entitled to the privileges of the sick role'.[28]

But if the mainstream refuses to put a name to your malaise or worse, suggests it might be related to your recent divorce or to overwork, alternative medicine is always there to offer unconditional and unquestioning acceptance of you, of the exclusively organic nature of your illness and of your money. Its practitioners can at worst be accomplices in sickening acts of deception. They can be relied upon to detect dysfunction in the healthiest subject or to recommend 'wellness care' on the rare occasions when they find nothing wrong. Even a beneficial bodily substance such as earwax, which plays a vital and health-promoting role in protecting the ear from infection, is deemed to be a disorder demanding the attentions of the neighbourhood ear candler. The slightly unpleasant gulp of mucous nearly everybody experiences, especially in the morning, is likewise a completely normal and benign physical process but is pathologised by alternative practitioners as 'post nasal drip' disorder which 'is a very embarrassing condition that causes irritation which can become very stressful'. Fortunately 'PND can now be treated safely, naturally and effectively', with SinusWars 3, a homeopathic product advertised with that familiar 85 per cent success rate.[29] PND? I told you I was ill!

It's not you, it's ME

Alternative practitioners are also keen to treat other problematic and so far medically unexplained conditions such as Chronic Fatigue Syndrome, or Myalgic Encephalomyelitis as it's known in the UK, for which there is no diagnostic test or known cure. Cases of CFS/ME are usually declared after the reporting by the sufferer of a number of symptoms, including extreme and constant tiredness lasting more than six months, generalised muscle pain and feeling ill after physical exertion. A diagnosis of CFS/ME is essentially one of exclusion, made when other possible causes are ruled out by laboratory and other investigations. It may follow a viral infection – both glandular fever and the herpes virus have been proposed as triggers – other illness, head injury or traumatic life event. It can last for months or years. The patient support group Action for ME says there are 240,000 CFS/ME suffers in the UK, most of whom are women between the ages of thirty and fifty.[30] Twenty-five per cent of sufferers are disabled by the illness to the extent they are unable to undertake minimal physical activity and 5 per cent are bed bound.[31] Whatever may cause or contribute to the condition, there is no question that CFS/ME symptoms are real and that they cause genuine suffering and consign many of those afflicted to a wretched existence.

Alternative practitioners advocate no single therapy. Instead they propose an astonishing variety of treatments, including special diets and supplements, massage, chelation therapy, cranial osteopathy, vitamin injections, acupuncture, homeopathy and the rest. Indeed it is hard to find an alternative practitioner who does not offer treatment for CFS/ME, despite there being no evidence that any of their therapies work.[32] And because mainstream medicine offers relatively little to people with this illness, sufferers frequently try one alternative therapy after another, spending thousands of pounds to no apparent benefit.

Florence Nightingale, bedridden for ten years after returning from the Crimean War; her birthday is now national ME awareness day

The British author Rachel Anderson has written a novel, *This Strange New Life*, based on her experience of having two sons who developed chronic fatigue in early adulthood. In a *Woman's Hour* interview she described her 'tremendous vulnerability to quackery' while they were ill. 'You spend your effort following up completely loopy things which are no use at all. My feeling of failure is that I didn't protect my sons enough from all the bizarre things that came up, like colour therapy.'[33] When television presenter Esther Rantzen's daughter Emily's ME became public, her family was similarly bombarded. 'The quacks had swept in,' she said. 'They wrote to us about a thousand different "infallible" cures: cold water baths; aloe vera drinks; oxygen at night; dowsing; feng shui; vitamin transfusions; magic crystals. We were advised to put Emily on a sugar-free, wheat-free diet. We were visited by a "white witch". I was told to rearrange all our furniture along the ley lines in the earth, to pick up mystic vibrations. We were sent copper bracelets and amulets, and a dozen self-help books.'[34]

It doesn't help that mainstream medicine and alternative practitioners, together with CFS/ME sufferers themselves, have widely differing views of the condition's nature, cause and perpetuating factors. Orthodox medicine tends to view CFS in terms of what it calls the biopsychosocial model, which involves an acknowledgement of a psychological or socially generated component to the illness. This is rejected by many sufferers – known as the ME lobby or ME community – who perceive a trivialisation of their illness and an implicit accusation that the condition must be 'all in the mind' or that they are malingerers or hypochondriacs. They believe that CFS/ME is a neurological illness probably caused by an immune or genetic disorder and that any psychological aspects are secondary. They are opposed to any doctors, particularly those they identify as members of the 'psychiatric lobby', who believe there are psychological and/or social aspects to either the development of the condition or its perpetuation.

The ME community also objects to what it sees as the bias in current CFS/ME research, complaining that 'the biopsychosocial model dominates the canvas in terms of research funding and exposure in professional journals instead of being a small part of the overall clinical and scientific picture', and urging government and research bodies to prioritise exclusively biological investigations of the condition.[35] This however is in contrast with a review published in *The Lancet* that said 'the use of a psychological CFS model does not preclude neurobiological components' and agreed that more research is needed into fatigue amongst CFS/ME sufferers and the population as a whole in order to understand it better.[36] Immunologist Professor Anthony Pinching, for one, has bemoaned the lack of investment and interest in solutions for patients with CFS/ME in comparison with those who have HIV/AIDS, saying that both conditions 'have shown strongly the need for a balance between the art and the science of medicine, and for continuity of care'.[37] What is remarkable here is that mainstream medicine's approach could be said to be considerably more 'holistic' than that of many ME patient advocates, who are determined to focus exclusively on the body as if the mind or social environment have no role to play in either the nature, course or experience of this or any other illness.

Treatments such as the psychotherapy method Cognitive Behaviour Therapy (CBT) and Graded Exercise Therapy (GET), in which sufferers gradually increase their level of physical activity, have been shown to be the most effective form of help.[38] [39] As a result of this the CBT and GET approach spearheads the latest UK National Institute for Health and Clinical Excellence (NICE) guidelines for the treatment of the condition. These guidelines are designed to be put into practice by the UK National Health Service's twelve new regional ME centres, established at a cost of over £8 million. But the NICE guidelines received an overwhelmingly hostile response from the ME lobby, who say that some patients, especially those with the most severe manifestations of the condition, are adversely affected by psychologically-based treatments. They were also disappointed to find that the new guidelines did not define the condition as a neurological illness.[40]

Feelings on the matter run high. Professor Simon Wessely of King's College London has been vilified by the ME lobby for pioneering what they call the 'Wessely School' approach, which includes psychologically-based interventions like CBT. Simon

Wessely set up the first NHS clinic for people with CFS/ME at King's College Hospital in London and he has been associated with research that has shown that while the condition may be triggered by an organic condition, psychological and social factors need to be addressed in order to facilitate rehabilitation. Professor Wessely and other researchers have been abused and intimidated because their work has been unpopular amongst powerful patient groups.[41] [42]

A recent *British Medical Journal* editorial agreed that 'the history of this field has been littered with miscommunications and misunderstandings,' but was more positive about the future for people with CFS/ME, concluding that, with its new treatment guidelines, 'NICE has forged a remarkable consensus and created a unique opportunity for us all to work together to provide the right care for the right patients at the right time.'[43]

Everybody hurts

In the late 1980s rheumatologist Dr Frederick Wolfe, director of the US Arthritis Research Center Foundation, noticed an increasing number of patients coming to him complaining of muscle pain and tiredness but with no observable physical signs of illness. When the patients were examined there were no detectable indications of muscle inflammation and laboratory tests revealed nothing abnormal. Dr Wolfe decided to recruit twenty colleagues to discuss the phenomenon, and soon after the term 'fibromyalgia' was coined, the Greek *myo* meaning muscle, *algia* meaning pain with the Latin *fibro* to represent connective tissue. It was subsequently decided that fibromyalgia should be diagnosed when a doctor pressed hard on eighteen specific 'tender points' on the body and pain was felt at eleven or more.

In common with CFS/ME, there is still no objective test to confirm the diagnosis of fibromyalgia. Some changes in brain chemistry have been noted in sufferers but it is unclear whether these are the result of the changes to lifestyle and activity, which come about when someone has the condition, or a sign that the disease has a physical origin. It has also been suggested that those with fibromyalgia might simply be more sensitive to pain than other people.[44]

There are now an estimated ten million fibromyalgia sufferers in the US, most of them women, who describe a mix of symptoms including joint pain, fatigue and decreased mental functioning they sometimes call 'fibro fog'.[45] The 'tender point' diagnostic method was backed by the American College of Rheumatology in the hope that it would enable researchers to find out more about this mysterious condition and how it could be treated. But there is still no medical consensus on fibromyalgia or whether it even exists. Many believe that its origin is predominantly psychological. Ten years after the term was created, Nortin Hadler, rheumatologist and professor of medicine at the University of North Carolina, has suggested that the lack of objective evidence to confirm its existence makes fibromyalgia essentially a condition generated by doctors themselves. He wrote that fibromyalgia was 'crafted to be consonant with our concept of disease and . . . nurtured by the rules of the physician–patient contact'. Fibromyalgia was not a syndrome, he argued, it was a system of belief.[46]

In recent years Dr Frederick Wolfe has been reported as wishing he could make the fibromyalgia diagnosis, which he helped create, disappear. He told the writer and doctor Jerome Groopman that 'for a moment in time, we thought we had discovered a new physical disease, but it was the Emperor's new clothes. When we started out, in the Eighties, we saw patients going from doctor to doctor with pain. We believed that by telling them they had fibromyalgia we reduced stress and reduced medical utilisation. This idea, a great humane idea that we can interpret their distress as fibromyalgia and help them – it didn't turn out that way. My view now is that we are creating an illness rather than curing one'.[47]

Undaunted by there still being no way medically to confirm a diagnosis of fibromyalgia, several large pharmaceutical companies, including Pfizer, Eli Lilly, Forest Laboratories and Wyeth are currently in a race to be the first to receive FDA approval for a fibromyalgia drug treatment, a market until now dominated by antidepressants and painkillers. FDA approval could lead to the recruitment of millions of new consumers but would also have an important role in confirming the disorder's legitimacy.[48]

Lyme theme

As well as inventing fad diseases, CAM practitioners have taken to making false diagnoses of real conditions. The story of Lyme disease shows how easily this can happen, to the point where thousands of people worldwide now wrongly believe they are in the grip of a serious infection.

In the mid 1970s the American rheumatologist Dr Allen Steere encountered a patient suffering from a collection of symptoms that no doctor had been able to explain: headaches, severe joint pain and fatigue. Steere's research led him to a number of other people, including children, with similar symptoms. He found that these cases occurred most often in summer, in people living in heavily wooded areas and who had been previously bitten by a deer tick. He named the condition Lyme disease after the Connecticut town, Old Lyme, near which it was discovered. He went on to find that its symptoms depended on the location in the body of the infection and that antibiotics were an effective treatment.

Despite initial scepticism, the existence of Lyme disease was subsequently confirmed by a number of other researchers. It was found to be caused by a spiral-shaped bacterium, *borrelia burgdorferi*, transmitted by the deer tick. Once described, there followed, in the ten years between 1982 and 1992, almost 50,000 reported cases of Lyme disease, which is now also known as borreliosis. Incidence has risen every year to the point where there are now around 23,000 new cases in America every year, mostly in the east.[49]

In Britain the Health Protection Agency puts the number of infections between one and two thousand per year, with the majority occuring in deer-friendly wooded areas like the New Forest, the Lake District and the Scottish Highlands.[50] [51] No one knows exactly why the incidence is increasing. One suggestion is that the bacteria which cause the disease have existed for centuries, but that it is now, with the increase in the deer population, global warming with its consequent mild winters and the popularity of country pastimes like rambling, more readily transmitted to humans.[52]

Chronic borreliosis or 'post-Lyme disease syndrome' has also been described, with symptoms, specifically Lyme arthritis, that may persist for years. It is thought that this can take hold if the initial infection was not treated effectively. And since fatigue is one of the later symptoms of Lyme disease, it has been seized upon by those with both CFS/ME and fibromyalgia as an explanation for their illness, who argue that it can be present even if blood tests are negative.[53] But Dr Darrel Ho-Yen, head of microbiology at the Raigmore Hospital in Inverness and an expert in both Lyme disease and ME, believes the link is very rare. He says 'in the Highlands of Scotland we have the greatest tick populations and it has been my routine in the investigation of CFS patients to have them tested for LD. In this large number of patients who have had very significant exposure to ticks, the number of ME patients who have had LD as the cause of their illness is around 5 per cent.'[54] There has also been a move to link Lyme disease with autism. In America the Lyme Induced Autism Association claims that up to 90 per cent of children with autism are infected with *borrelia* bacteria.[55]

Once again, the internet proves the most fertile breeding ground of all. In the past few years it has hosted an explosion of information about Lyme disease, much of which has been shown to be inaccurate.[56] This, together with the proliferation of unorthodox and unvalidated testing methods which frequently produce false positives, has led to an over-diagnosis of Lyme disease, according to the UK Health Protection Agency. An investigation in 2006 by Professor Brian L Duerden, concluded that 'none of the unorthodox diagnostic tests purported to support the diagnosis of Lyme Borreliosis or other infections in chronic fatigue syndrome are validated and should not form the basis of any medical diagnosis or treatment prescription or recommendation'.[57] [58] The Infectious Diseases Society of America, in its latest Lyme disease treatment guidelines, agrees that the diagnosis should not be made unless the patient has had objective evidence of *borrelia burgdorferi* infection, using tests 'performed by well-qualified and reputable laboratories that use recommended and appropriately validated testing methods and interpretive criteria'.[59] In England and Wales these are carried out by the Lyme Detection Unit based at Southampton General Hospital.[60]

Back in America though, one specialist was becoming increasingly wary of both the spiralling incidence figures and the authenticity of the by now widespread phenomenon of Lyme disease. It was

Dr Allen Steere, who had discovered it in the first place. He decided to analyse the diagnoses and test results of a group of 788 patients referred to his clinic and found that only 180 had active Lyme disease. A further 156 showed evidence of having had the infection in the past, but everyone else, amounting to 57 per cent of the group, did not and never had had it. Instead, Steere found, most of these patients had symptoms consistent with either chronic fatigue syndrome, fibromyalgia or some kind of rheumatic disease. Before referral to the clinic more than half of the patients had been given antibiotics, which were mostly ineffective. 'The most common reason for lack of response to antibiotic therapy was misdiagnosis,' Steere concluded. Forty-five per cent of those in whom no evidence of Lyme disease could be found had received positive diagnoses from other laboratories using the non-approved and unreliable tests slated by Professor Duerden.[61]

These and other findings led Dr Steere to decide to speak out against over-diagnosis of Lyme disease and to highlight the risks of unnecessary long-term antibiotic use. Lyme disease patient groups responded with vocal opposition, to the point where Allen Steere has been in fear of his life. In common with Simon Wessely, he is the focus of virulent criticism from a new brand of patient advocacy group that has emerged to champion the rights of sufferers of medically unexplained conditions. Such groups create what David Grann of the *New York Times* has described as a 'parallel universe' which exists alongside the scientific establishment. Rather than lobbying for better treatment or increased research funds – as HIV/AIDS and women's health campaigners might have done in the past – the Lyme Disease Foundation in America 'seemed to go even further, trying to discredit the existing research and come up with its own,' says Grann. The organisation has 'created their own scientific publication, *The Journal for Spirochetal and Tick-borne Diseases*; organised its own scientific conferences; financed its own research and trumpeted its own medical experts with their own treatments'.[62] These groups are part of what historian Edward Shorter calls 'the subculture of illness' with its 'patients' support groups, claques of sympathetic (and very pricey) physicians, and periodic media concern'.[63]

The cultural critic Elaine Showalter points to similar developments in relation to other controversial conditions. She says 'what's really troubling about so many of these syndromes is the way they almost immediately escalate to conspiracy theories. As soon as people begin to feel convinced that there is some external cause for what they're feeling, they then believe that – since they can't produce any evidence – there must be a conspiracy to conceal that evidence and to cover it up. So they're in a kind of oppositional relationship to government, to medicine, to various forms of authority. And also they move more and more into a kind of superstitious environment where they will not accept the kinds of proof that science is now able to offer us'.[64] It is in exactly this superstitious environment that alternative medicine thrives.

But the controversy over chronic Lyme disease persists. Research and clinical experience have established that Lyme disease is extremely difficult to contract, even from a deer tick in a Lyme infested area, and that it can be treated immediately and effectively with a short course of antibiotics. In one American Lyme disease expert's view Lyme disease 'although a problem, is not nearly as big a problem as most people think. The bigger epidemic is Lyme anxiety'.[65]

Stone free

If you feel unwell it is good to have something to show for it. And if you are impressed by the mucky deposits in an ear candle you will be thrilled by the contents of the lavatory following the ingestion of one of the liver or gall bladder flushes popularised by Hulda Regehr Clark and others. These are claimed to be 'a quick way of flushing toxins, fat, sludge and small gallstones out of the liver and gall bladder'.[66] Clark claims that 'cleansing the liver of gallstones dramatically improves digestion, which is the basis of your whole health. You can expect your allergies to disappear, too, more with each cleanse you do! Incredibly, it also eliminates shoulder, upper arm, and upper back pain. You have more energy and increased sense of well-being.'[67] These organ flushes usually involve drinking a mixture of olive oil, grapefruit or lemon juice, Epsom salts (magnesium sulphate) and water. A few hours later you will pass the resultant 'stones', which Clark recommends you should collect in a kitchen colander placed over the lavatory. They'll look something like this:

Gall stones or soap stones? Products of a gallbladder flush

Are these really gallstones? A New Zealand case reported in *The Lancet* in 2005 provides a more credible explanation. A 40-year-old woman attended the Waikato Hospital in Hamilton with a three-month history of abdominal pain after eating fatty food. She had recently followed a liver-cleansing regime, using a flush recipe similar to the one above, on the advice of her herbalist. She collected the resultant semi-solid 'stones' and kept them safe in her freezer before presenting the collection illustrated above at the clinic.

Microscopic examination of the stones showed however that they were not gallstones, but pellets of a soap-like substance created by the mixing of olive oil and lemon juice with the stomach's own digestive enzymes. This concoction brought about saponification, whereby fats are made into soap, leading researchers to conclude that flushing regimes for expelling gallstones are a myth and the source of misleading claims.[68] Unfortunately this has done little to arrest the flushing movement, which is still promoted by Hulda Clark and in books like *The Liver and Gallbladder Miracle Cleanse*, by Andreas Moritz, who claims to be one of the 'world's leading experts on Integrative Medicine' and is the author of *Cancer Is Not A Disease! It's A Survival Mechanism*.[69]

Back in New Zealand meanwhile, an abdominal ultrasound revealed that the woman did indeed have a number of small gallstones, but that they had remained in her gallbladder. Once removed by surgery they would have never bothered her again.

Plaque attack

'When I eliminated 40ft of foul smelling rubbery slime, my life was changed! I am sooooo glad it is no longer inside me! God knows what kinds of awful things it must have been doing in my body, and no wonder I've been feeling sluggish and tired!'[70] So reads a testimonial to the positive power of 'colon cleansing' and how it rids the body of a health-sapping substance known as mucoid plaque, which can supposedly lurk in the digestive system for years.

The term mucoid plaque was invented by American naturopath Richard Anderson, who visualises our insides as something akin to Dante's Circles of Hell. In answer to the question 'Should I colon cleanse?' he replies 'the colon is a sewage system, but by neglect and abuse it becomes a cesspool. When it is clean and normal, we are well and happy; let it stagnate, and it will distill the poisons of decay, fermentation and putrefaction into the blood, poisoning the brain and nervous system so that we become mentally depressed and irritable; it will poison the heart so that we are weak and listless; poisons the lungs so that the breath is foul; poisons the digestive organs so that we are distressed and bloated; and poisons the blood so that the skin is sallow and unhealthy'.[71] I reckon that's a 'yes'.

Mucoid plaque has never been identified by anyone in mainstream medicine, with pathologist Dr Edward Uthman typical in asserting 'I have seen several thousand intestinal biopsies and have never seen any "mucoid plaque." This is a complete fabrication with no anatomic basis.'[72] In common with so many altmed practitioners confronted with such statements, Anderson explains the doubt about mucoid plaque by saying that 'doctors do not know what they are looking for when they say they can't see it'.[73]

There's no disputing the reality, however, of the substance that emerges from the body after a colon cleanse. Cleansing literature veritably overflows with photographs of these trophy motions. But, just like the liver flush, they turn out to be artefacts generated by the treatment itself. Most colon cleansing products contain a combination of psyllium and bentonite. According to the Natural Medicines Database, the plant psyllium 'forms a mucilaginous mass when mixed with water and has a bulk laxative effect'.[74] Bentonite is a type of clay that expands when wet – it can absorb seven to ten times its own weight in water and will swell up to eighteen times its dry volume. As well as being used in pottery, it is employed in civil engineering and as a 'clumping agent' in cat litter.[75] Once swallowed, psyllium and bentonite clay combine to make a soft, rubbery cast of the intestines, which is then passed and, all too frequently, photographed.

According to Dr Natura, a popular cleansing website, everyone needs internal cleansing 'unless you've been a vegetarian since birth and lived on a pristine island away from pollution growing your own food'. Their considered advice is to buy their products in bulk: 'Remember, it took years or even decades to accumulate all the toxins in your body, so it's impossible to get rid of everything in a few days or even a couple of weeks . . . With our 50% discount your best choice is buying three or more packs.' And there's still no stemming the flood of testimonials. 'I've been on the Colonix cleanse for three weeks so far, beginning my fourth week and I have at least two to three very healthy bowel movements a day. It's like a hippopotamus is using my bathroom, soon I may have to buy a bigger toilet!' Colon cleansing is also good for despatching those ubiquitous parasites. 'At first I saw only tiny little creatures and then a few days later the Loch Ness Monster arrived!'[76]

Toxic waste

The detoxification craze, which includes treatments for a multitude of disorders as well as endless television programmes and newspaper articles, shows the appeal of the purge to be as powerful as ever. Professor Edzard Ernst has described the ancient theory of autoin-toxication, which is the belief that intestinal waste can poison the body and is a major contributor to many, if not all, diseases. He writes 'in the 19th century, it was the ruling doctrine of medicine and led to "colonic quackery" in various guises . . . When it became clear that the scientific rationale was wrong and colonic irrigation was not merely useless but potentially dangerous, it was exposed as quackery and subsequently went into a decline.' Ernst notes that 'today we are witnessing a resurgence of colonic irrigation based on little less than the old bogus claims and the impressive power of vested interests.' These same bogus claims underpin the practice of the detox.[77]

The detox industry is believed to be worth tens of millions of pounds in the UK and is spearheaded by self-styled 'nutritionists', most of whom, in common with 'Dr' Gillian McKeith, have no training or qualifications beyond an unaccredited correspondence course PhD in 'holistic nutrition'. Gillian McKeith agreed to drop the 'Dr' moniker from her advertising after a complaint was made to the Advertising Standards Authority.[78] The title was finally purged from her website, books and supplements in 2008. Each New Year sees a massive marketing campaign to sell the idea of the detox in every pharmacy, supermarket and women's magazine. Our guilt at seasonal over-indulgence in food and alcohol, it seems, makes us welcome the unpleasantness and expense of the detox, so as to emerge puri-fied and, we hope, several pounds lighter.[79]

The first targets for banishment, always described as if they are inherently evil, are all white foods: white flour, white rice, white sugar, dairy products and salt. Don't imagine you might hang on to a spoonful of that nursery favourite, macaroni cheese. *Observer* newspaper nutritionist Dr John Briffa counsels pithily 'from a nutri-tional perspective, I have to say that a dish made out of refined wheat and cheese sauce is utter shite'.[80]

As well as a radical change in diet, detox usually involves consuming an assortment of vitamins, minerals and laxatives, washed down with gallons of water or, if you're lucky, some herbal tea. You can help the process along with a 'detoxifying body wrap' like Shape Changers. First you are daubed with 'natural sea clay' and then trussed tightly in bandages from the neck down: 'As the toxins are drawn out the soft tissue is compressed and compacted as the skin regains elasticity by being detoxified thereby creating a smoother and firmer muscle base over which the soft tissue lies,' claim the manufacturers.[81] It might be less messy if you 'try the easy way to detox with the best quality, totally natural Detoxing Footpads . . . simply place one on your foot over night and wake feeling refreshed in the morning'. Crystal Spring Natural Detoxing Footpads are teabag-like 'sachets of bamboo and tree extract laced with powdered tourmaline and loquat leaf' that should be stuck on your feet every night for at least a week. A box of fourteen sachets costs £24.95.[82] In Australia a ten-day Lemonade detox diet is being marketed to children as young as six: developed thirty years ago by self-styled 'Master Cleanser' and naturopath Stanley Burroughs, this involves drinking nothing but lemon juice mixed with water, cayenne and a substance called Madal Bal Natural Tree Syrup.[83]

Carol Vorderman's 28-day detox is said to 'do wonders for the way you look, the way you feel and your ability to cope with stress'. It's a low-cal vegan regime which as well as exorcising the white stuff, excludes all meat, fish, eggs, alcohol, fizzy drinks and ready meals. If Carol is to be believed, everyone needs a detox. 'You probably don't realise that you are bombarded by toxins every day. These come from pollutants in the air, cigarette smoke, exhaust fumes, detergents, household chemical and toxic metals in the environment such as mercury and lead. And then there's the cocktail of pesticide residues and artificial additives in your food as well as the toxins found in alcohol, caffeine, drugs and medications.'[84]

It is a shame that Carol Vorderman, with her superb mathematical skills, appears to misunderstand the principle believed to be first formulated by the alchemist and physician Paracelsus in the sixteenth century: 'all things are poison and nothing is without poison, only the dose permits something not to be poisonous'. These days this is often shortened to 'the dose makes the poison'. So while water is essential for life, an excess can cause water intoxication, known as hyponatraemia, a potentially fatal brain condition. Caffeine is not a poison unless taken in overdose when it can cause breathing difficulties and convulsions. And the vitamin A that Vorderman recommends in small amounts as a supplement can cause orange skin, blurred vision, nausea and (in rare cases) growth retardation, hair loss, enlarged spleen, liver damage or even death if you take too much of it.[85]

Enthusiasts invariably describe unpleasant effects, including body odour, bad breath and spots, alleged to be a sign that the toxins are leaving the body. But what these disparate 'treatments' have in common is that there is never any identification of what these emerging toxic substances might be, and certainly no evidence that these unnamed materials have actually departed. Instead, as ever, feeling worse is a positive sign and not an indication that the detox regime is doing you no good at all. Moreover you should see it as a prompt to purchase yet another detox product, like Boots' own Detox Body Freshener tablets.[86] Once you're done with the detox you can protect yourself against further contamination with a £30 face cream from Clarins, Expertise 3P, claimed to be able to shield your skin 'from the accelerated ageing effects of all indoor and outdoor air pollution' as well as 'the effects of Artificial Electro-magnetic Waves'.[87]

Signs of a backlash against the detox industry came in 2006 when a group of British scientists and doctors joined forces to try to dissuade people from the practice, pointing out, amongst other things, that the human body has its own detoxifying mechanisms. 'The gut prevents bacteria and many toxins from entering the body. When harmful chemicals do enter the body the liver acts as an extraordinary chemical factory, usually combining them with its own chemicals to make a water-soluble compound that can be excreted by the kidneys. The body thus detoxifies itself. The body is re-hydrated with ordinary tap water. It is refreshed with a good night's sleep.' They said detox products and rituals 'waste money and sow confusion about how our bodies, nutrition and chemistry actually work'.[88]

As part of BBC 2's *The Truth About Food* series, ten women (self-confessed 'party animals') were taken to a country cottage retreat for ten days 'to see if a detox diet could recharge their internal batteries'. This (admittedly small) group was split into two and 'half the girls were put on a balanced diet, including red meat, alcohol, coffee and tea, pasta, bread, chocolate and crisps (in moderation), with the remainder following a strict vice-free diet'. This meant no meat or dairy foods, constant juicing, herbal teas and the consumption of a list of supposedly toxin-free foods, devised by three 'nutritionists'. When the week was over the women were tested to measure kidney and liver function as well as the levels of antioxidants and aluminium in their blood, all of which would have indicated what impact the detox had on the body. There were no differences between the two groups.[89]

Illness brokers

There is, mercifully, increased awareness in both the medical profession and the media of the rise of what's known as disease mongering, the selling of sickness that widens the boundaries of illness in order to expand the market for doctors and drug companies. And there has been widespread criticism of pharmaceutical industry-funded disease-awareness campaigns, often 'designed to sell drugs rather than to illuminate or to inform or educate about the prevention of illness or the maintenance of health'.[90]

But in direct contrast to what is often claimed, alternative and complementary medicine, with its facility for inventing, detecting and treating illnesses and conditions that orthodox medicine and even the pharmaceutical industry cannot identify, beats the mainstream hands down when it comes to the mongering of illness and disease. It's enough to make you sick.

Chapter 9

It Works for Me

It is worth restating the simple fact that there is no evidence that complementary and alternative medicine can cure any disease. In America the government-funded National Center for CAM has failed to prove any single method to be effective, despite a research budget of almost one billion dollars. In Europe and elsewhere, trial after trial finds CAM either (like homeopathy) to be no more effective than placebo, or, like chiropractic neck manipulation and some herbal remedies, to be positively dangerous. In terms of its evidence base, alternative medicine has, time and again, been well and truly busted.

But this has done nothing to halt its rise and rise, and questions of evidence have almost become a side issue in the debate about alternative medicine's value. Instead, a complex combination of superstition and market forces propels the integration of CAM into mainstream health care, regardless. So we have to go beyond expositions of hard evidence to ask why so many people turn to these discredited practices with their unsubstantiated claims. What is the attraction of alternative medicine and why do people return to its practitioners again and again?

'Cos you gotta have faith

What is in no doubt is that many people enjoy using alternative medicine and that they genuinely believe it works. It matters not that the chosen method or remedy is irrational, theoretically at odds with other alternative methods or remedies, or that it has been repeatedly disproven by scientific research. The behaviour of CAM users is not influenced by concerns about plausibility, consistency or evidence. In this sense alternative medicine might arguably amount to a faith-based practice, a religion in all but name. Its followers are encouraged to venerate the so-called 'life force' – variously named Qi energy, Innate Intelligence or the Breath of Life – which is manifest in the form of 'healing energies', and deemed to reside both in nature and within their own bodies. Disciples of the CAM faith must pay constant attention to these internal energies and to their own well-being in this cult of the self.

Followers may also continue to use alternative medicine regardless of any contrary information or even negative experiences they may have. They are what are known as 'true believers', a term coined by reformed fraudulent medium M Lamar Keene in the 1970s. The true believer, he says, is an otherwise sane individual who becomes 'so enamoured of a fantasy, an imposture, that even after it's exposed in the bright light of day he still clings to it – indeed clings to it all the harder.'[1] Once the burden of evidence is not required anything can simply be asserted. Say it is so and it is so. A dash of ritual leads both to a whole new alternative medicine speciality and product range.

The credulousness of the alternative medicine consumer is demonstrated spectacularly by the marketing spiel for Indigo Essences, 'a kind of first aid kit for feelings' invented by homeopath Ann Callaghan. She describes them as 'good vibes in a bottle'. To cut a long story short, she dips rocks in water, dilutes that water and sells it as an 'essence' for £8.95 for a 15ml bottle, or as a further diluted spray at £12.95 for a 100ml bottle. It's a questionable sort of parenting that entails squirting children with water from bottles labelled Confidence, Happy, Settle or Shine. And one can't help but feel concern for the psychological health of the child who is sprayed by their mother with a blast of Invisible Friend for 'when you are feeling lonely or scared and you wish someone was there to give you a big hug'.[2] Lest you think that Indigo Essences are at the extreme end of alternative medicine use, they have much in common with those homeopathic products branded Coldenza, Rheumatica or Teetha available from high street chemists. Whether it is sugar pills or water being marketed, the active ingredient is all in the name.

The manufacturing process may have sounded simple, but as so often in alternative medicine, the attendant ritual and communion with spiritual forces performed under the auspices of a priest-like figure, is how value is added. 'First,' says Ann Callaghan, 'I get an "intuition" to make an essence with a particular rock or two. Then I ask the rocks if this is true.' She used to use a dowsing pendulum to secure a reply, but these days just does a bit more intuiting. 'First I call in a special spirit team of essence making experts and we all plurk (play & work) together – this is called co-creating. I gather the things we need to make the essence. A crystal bowl, good clean water, things to make the room nice like candles and flowers and anything else that feels right. The spirit of the rock is healed and cleaned so that only its most pure and healing vibration goes into the water during the essence making process . . . Different rocks ask for different things. Some want to be put in bowls of water outside in the sun, some like the moon, some don't want to be put in water at all and just sit beside the water bowl . . . Sometimes I see energies going into the water. Sometimes I just feel things'. And sometimes she sets up a distribution network with her business partner Barbara to enable Indigo Essences to be marketed worldwide and sold in around eighty outlets in the UK.

What she wants

We know the prime users of alternative medicine worldwide – it's those middle-aged, middle-class, educated women with a high disposable income. The younger end of this group is also likely to take their children to naturopaths and cranial osteopaths, to avoid having them immunised and to medicate them with shop-bought homeopathic and herbal remedies. Alternative medicine offers these women a way to take control, to be remarkable in their day-to-day lives and to make them feel as if their needs as individuals are being attended to. It touches them, both physically and emotionally, at a point in mid-life when many women in our society say they are beginning to feel invisible. It tells them that they are unique, different from every other person and, importantly, more, much more, than the sum total of their symptoms. And as we shall see, it provides the kind of positive reinforcement than most adults could not reasonably expect from their closest friends, or even their partner. Indeed, 'nutritionist' Gillian McKeith offers a cereal bar which lists 'unconditional love' amongst its ingredients.[3]

Marketing executives have been quick to appreciate the strong appeal of CAM for women. For years beauty products have been promoted as being more medicinal than cosmetic, with their pharmaceutical-sounding ingredients like Boswelox™, pentapeptides and active liposomes. Alternative medicine returns the compliment with herbal, aromatherapy and even Ayurvedic ingredients now residing on every beauty counter. Aromatherapy is the foremost occupier of the no-man's land (literally – there is no male presence) between the cosmetics and the alternative medicine industries. Stuck for a Mother's Day gift? How about an 'Aromatherapy Pamper Day – perfect for mums!'[4] This, it is assumed, is what every woman must desire.

Flower remedy workshop

The writer Julie Burchill anatomises how 'pleasure, for women, has been pathologised in recent years'. She argues that the idea of 'pampering' in particular has become a sort of therapy in itself, and in typically forthright style rails against the political implications, describing how 'the scented "aromatherapy" candle is one of the foremost weapons in middlebrow England's dirty little war against feminism – its message being "Calm down, you hysterical cow, and stop complaining. You stink! And you're so ugly, don't let any man see you with the light on, he'll run a mile!"'[5] And if pleasure has been pathologised, then the reverse is also true – attending to one's pathology must be infinitely pleasurable. It is extraordinary what can be achieved by having a nice long soak in a bath infused with a squirt of 'Immune System Tonic Aromatherapy Bath Oil'. It contains 'melissa to counteract fatigue, tea-tree to stimulate the body's own defences, lavender to reduce anxiety, bergamot to lift the spirits and regain mental focus, sandalwood for courage and long life and extract of ginseng for physical rejuvenation'.[6]

Equally, you might enjoy a therapeutic massage, popular with have-it-all mothers like stressed out actor Tracy-Ann Oberman, who reveals all in the *Guardian*. "'There's no way I'm going to be able to relax," I warned Kylie, explaining that I had a chronic lower-back ache and shoulders so tense they were up to my ears. Ten minutes later I was snoring face down as those wonderful creams were massaged into my skin. I started to feel human again. "What bliss," I mused. "This is all about me, me, me."'[7] Many alternative treatments offer much pleasurable gain in exchange for minimal pain. As comedian Helen Lederer confirms, 'the great thing about reflexology is that you can feel all the veins and gristle being crunched, and at the end of it you feel like you've done a workout but all you've done is lie there'.[8]

The 'ultimate indulgent treatment' is said by its practitioners to be 'LaStone therapy', a massage involving the use of hot and cold stones claimed to increase circulation, metabolism and lymph function as well as to detoxify and boost the immune system, whatever that means. One user describes how the therapist 'starts by burning sage to clear negative energy – a practice borrowed from the treatment's Native American roots. I lie with my back on a layer of stones, mainly hot but with a few carefully positioned cold ones. Stones are placed on my body and forehead. Then Sarah massages me with the rocks, alternately hot to comfort and cold to invigorate. It is almost too hot and too cold to bear, but never quite'.[9] According to the LaStone practitioner in question, Sarah Kirby, this treatment 'enables your body to be healed, whilst at the same time allowing you to enter a sublime level of relaxation which you may never have experienced before.'[10]

As these accounts confirm, alternative medicine knows precisely how to make every user feel special. CAM says you are unique so your treatment needs to be carefully calibrated to reflect your individuality. In the name of 'treating the whole person' it wants to know all about *you* – there is no detail of your life too insignificant to be of interest to a CAM therapist. As one North London complementary medicine practice puts it 'each patient is unique and every homeopathic prescription is individual. Each of the remedies has a symptom picture and the one that matches you the best is the one which will be prescribed. Therefore the more information you can give about yourself in the consultation the better, this includes your temperament and moods, medical history including any recurrent illnesses and allergies, food likes and dislikes etc'.[11] What matters is *you*, not your illness symptoms or even whether you actually have any identifiable illness or symptoms. Alternative medicine 'treats patients, not data' because 'disease labels are generally unhelpful . . . what is important is the characteristic symptoms the individual is exhibiting, the unique picture that person is presenting.'[12] It is an alluring prospect. As Professor Raymond Tallis puts it: 'alternative medicine promises – groundlessly of course – an interpretation of illness that is both effective and integrated into the rest of your life. Because it has a kind of intuitive appeal, and personalises disease processes, it narrows the gap a little between the alien facts of your case and your immediate suffering.'[13]

The mainstream practitioner simply can not and will not offer such an amount of time or interest in the fine details of your life. In their book *Why Do People Get Ill?* Darian Leader and David Corfield remind us that the average duration of a consultation with a London GP is now the shortest in Britain, at between six and eight minutes. Elsewhere it's usually fewer than ten. On average, a patient speaks to their GP for just 23 seconds before being interrupted. Leader and Corfield suggest these statistics are only one aspect of the reason why so many turn to alternative practitioners. 'As the body has become more and more fragmented, and health services reduced to local applications of intervention and expertise, the individual has been lost in a vast maze of conflicting interests and misguided techniques. This has led, quite naturally, to an appeal to alternative forms of medicine that take the time to listen and to respect the particularity of the patient.'[14]

In a consultation with an alternative therapist the patient is able to define for themselves the nature of their 'particularity' so that the consultation can follow their own agenda, as GP Margaret McCartney has noted. She thinks 'the core values of care and continuity need to be valued in NHS primary care' and that 'it is scandalous that the one place on the NHS where patients are given the time and place they need is a homeopathic hospital. Why should one have to leave science behind in order to gain the time and perception of care which should be accessible to all?'[15]

Whilst CAM does well in terms of listening and the time it allows its patients, making it seem more patient-centred than orthodox medicine, the reality is more complicated. The 'doctor knows best' medical paternalism of yesteryear is increasingly absent in conventional health care, yet is thriving in alternative medicine. As altmed sceptic and journalist Dr Ben Goldacre points out, these days 'transparent modern medics often say "I don't know what the cause of your problem is. This might make it better, but it might not, and it might have these side effects." They sometimes follow this with: "What do you think?"' But here comes the alternative therapist, 'who understands your problems whatever they are, who is privately employed and has time to listen, who has an answer and who gives a complicated (often wilfully obscure but always authoritative) explanation of what is going on, maintaining the power imbalance in the therapeutic relationship with his or her exclusive access to arcane knowledge. If that's not old-fashioned medical paternalism, I don't know what is.'[16] It is an abiding paradox that alternative medicine is used most keenly by the generation of women who, in the form of the women's move-ment of the 1970s and 1980s, asserted it was 'our bodies, our lives, our right to decide' and rejected paternalistic medicine in the delivery room and beyond. Yet these same women now want to be told what to do by a shaman.

But CAM conceals its directive paternalism by giving users a limited sense of empowerment via the practice of self-prescribing. Users are granted the minimal knowledge but the extensive means to be their very own wise women, with a practice founded on nothing but a smattering of superstition and a large helping of female intuition. We can obtain a complete CAM armamentarium from health food shops and high street chemists like Boots with their extensive 'Alternatives' range of remedies which 'can help give you a sense of control over your own health care'. The majority of homeopathy users prescribe both for themselves and for members of their families and it has been estimated that up to 20 per cent of Britons have bought homeopathic medicines over the counter, at a pharmacist or health food shop.[17]

The attractions of self-prescribing are that it is not an exact science and that there's no need to undertake years of wearisome medical education. All-purpose sets like the Helios Double Helix Homeopathic Remedy Kits contain up to thirty-six different reme-dies in 'sucrose pillule' (also known as a sugar pill) form. 'Each kit comes with its own comprehensive, self-prescribing guide to help you in your selection of the most appropriate remedy'.[18] Those who favour Bach flower remedies can find out how in a booklet 'You and the Bach Original Flower Remedies' where they will learn 'if you have selected more than seven remedies cut down your choice by concentrating on how you feel now. If a feeling isn't affecting you today you don't need a remedy for it'. If you are still uncertain 'it can be a good idea to ask close friends what they think you need – they may see things that you have missed'.[19]

Wet blanket

In common with many organised religions, alternative medicine also understands how much we like to be told how to behave, gratifying this instinct by providing a set of rules for living and a collection of consequent virtues. Instead of those burdensome ancient holy virtues of charity, abstinence, humility and the like, CAM offers a variety of more easily achievable twenty-first century virtues, a classic example of which is the virtue of full hydration. To witness how an alternative health practice can be welcomed and fully integrated into everyday life, look no further than the ubiquitous exhortation that we need to drink eight glasses of water every day in order to maintain health and well-being. As each glass is supposed to contain 8oz of water this routine is commonly abbreviated to '8x8' and amounts to a daily consumption of around two litres or half a gallon of water. But this is just for starters – model and actor Jerry Hall claims she drinks six litres of water a day.[20]

In veritably biblical style, 8x8 has been described as the first commandment of good health and is supported by a number of excessive claims. More than half of the population is permanently dehydrated, we are told; you can be dangerously dehydrated without even knowing it; and you can only properly rehydrate with pure water and not by drinking other drinks like tea, coffee, soft drinks or alcohol.[21] A little like the 'silent killer' that is the chiropractic 'subluxation', it is suggested that dehydration can sneak up on us unawares. We are warned that once we become thirsty, we're already starting to get dehydrated and that 'by the time we decide it's time to stop and get a bottle of water, we're already approaching a level of dehydration that would be physiologically compromising'.[22]

8x8 is alleged to help improve energy levels, as well as mental and physical performance. It is also meant to keep skin looking healthy and promote weight loss. It has even been claimed that water can cure diseases including arthritis, asthma, autoimmune diseases, back pain, colitis, diabetes, heartburn, high blood cholesterol, high blood pressure and migraine. Dr Fereydoon Batmanghelidj, author of *Your Body's Many Cries for Water* and the foremost champion of water's curative powers died in 2004, but left a 'message to the world'. He told us 'You are not sick, you are thirsty. Don't treat thirst with medication'.[23] Dr Batmangehelidj was reported as claiming that his ideas represented a 'paradigm shift' and that he had to publish his book himself so that his findings would not be suppressed.[24]

Such widespread promotion of 8x8 is of course a sales opportunity not lost on the bottled drinks industry. The volume of bottled water sold in the UK has reached almost one billion litres per year with water predicted to represent 21 per cent of soft drink sales by 2008.[25] According to research by drinks manufacturer Britvic, bottled water now out-sells cola in London.[26] Those who drink water, they say, are following what the industry calls the 'better-for-you-trend' and are keen to remind us of the importance of drinking 'six to eight glasses of fluid per day to help our body function at its best'.[27]

Physiologist Dr Heinz Valtin of the Dartmouth Medical School in New Hampshire decided to investigate the origins and evidence for the putative benefits of 8x8 after noticing how it had become 'perfectly acceptable to sip water anywhere, such as during lectures, seminars and conferences. A colleague has told me that he estimates that something like 75 per cent of his students carry bottles of water . . . for some, the bottle has even become a security blanket: recently as I listened to a postdoctoral fellow presenting a seminar, I observed that whenever his flow of words stopped momentarily, while he contemplated the next sentence, he would, seemingly unconsciously, pick up a bottle of water from the table, unscrew its top and replace it, without ever taking a sip'.

After reviewing a collection of scientific studies and surveys of food and fluid intake on thousands of adults, Dr Valtin could find absolutely no evidence in support of 8x8, nor any evidence that the majority of the population was suffering from dehydration. He concluded that the advice might have originally stemmed from a misreading of guidelines issued as long ago as 1945 by the US Food and Nutrition Board. These stated that 'a suitable allowance of water for adults is 2.5 litres daily in most instances. An ordinary standard for diverse persons is one millilitre for each calorie of food. Most of this quantity is contained in prepared foods.'[28] For some reason the last sentence was either forgotten or ignored, leading, over the years, to the idea that an *extra* eight glasses of water should be drunk every day.

In fact the water contained in food represents as much as one half of an average daily intake. As well as being the principal ingredient of fruit and veg – broccoli is around 90 per cent water – it is present in most other foods. A steak is 65 per cent water, bread is 35 per cent water and even hard cheese contains 38 per cent.[29] [30] Dr Valtin also points to research that shows that other drinks, including caffeinated ones like coffee, tea, soft drinks and colas, and to a lesser extent dilute alcoholic beverages such as beer in moderation can be included in total fluid consumption.[31] Even though some of these drinks may be mildly diuretic, they still represent a net gain to one's total liquid intake. And drinking tea could even be better than drinking water. According to one recent review, tea replaces fluids and contains antioxidants thus offering a potential double benefit, possibly protecting against heart disease and some cancers.[32]

Heinz Valtin emphasised that his conclusions apply to healthy adults in a temperate climate, leading a largely sedentary life. Those doing strenuous exercise, who live in hot climates or who have specific medical conditions, may need more fluids. Drinking an excess of water is not without risk – it can cause water intoxication with its consequent mental confusion, seizures, and possible death. Such conditions are being reported with increasing frequency in endurance athletes, military recruits and people using recreational drugs that cause extreme thirst.[33]

Wishful thinking

Much of CAM's appeal derives from users' subjective experience of the effectiveness of their chosen therapy. In his essay 'Why bogus therapies often seem to work' the late Canadian psychologist Barry Beyerstein showed how temporary mood improvement can easily be confused with cure. 'Alternative healers often have forceful, charismatic personalities. To the extent that patients are swept up by the messianic aspects of "alternative medicine", psychological uplift may ensue.'

But feeling better is not the same as being better. In a dramatic example of this phenomenon Gambian president Yahya Jammeh has, he claims, created a cure for HIV/AIDS consisting of herbal body pastes and other preparations. Like many of the pioneers of alternative medicine, he says the treatment was revealed to him by his ancestors in a dream. In a move which has brought criticism from both the World Health Organisation and the United Nations Development Programme, patients are instructed to stop taking their anti-retroviral drugs before treatment. In Jammeh's mind there is no question that the cure works. 'My method is foolproof,' he told reporters. 'Mine is not an argument, it is a proof. It's a declaration. I can cure AIDS and I will.'[34] Crucially, his patients, who are treated for free, believe themselves to be completely cured. One man, a 54-year-old who has been HIV positive since 1996, said afterwards 'It feels as if the President took the pain out of my body . . . I am cured at this moment . . . as I stand before you I can honestly tell you I have ceased to have any HIV symptoms'.[35] There has been no independent confirmation of Jammeh's claims and a UN envoy who questioned them was expelled from the country.[36]

Barry Beyerstein also described how people who hedge their bets often credit the wrong thing, as we saw in relation to cancer treatment. 'If improvement occurs after someone has had both "alternative" and science based treatment, the fringe practice often gets a disproportionate share of the credit.'[37] If we feel better in the same time frame as the administration of an alternative remedy, it is natural to associate the two: it appeals to the human desire to derive order and meaning from any combination of events.

Writer Karl Sabbagh illustrates how this works in the mind of the alternative medicine user. He says 'we all have a tendency to look for patterns in the world and make links where none exist. If your constipation disappears shortly after someone has inserted a needle with great ceremony in a very specific part of your earlobe, few of us would doubt that one causes the other, even though constipation around the world gets better every hour with little or no insertion of needles into ears.'[38] This time-based association leads to the logical fallacy famously expressed in Latin as *post hoc ergo propter hoc* which means 'after this, therefore because of this'. Once an association has been forged in our minds, belief follows.

The power of belief is demonstrably apparent in Applied Kinesiology, a diagnostic method using muscle testing invented in the 1960s by an American chiropractor, George J Goodheart. AK, it is claimed, can be used to diagnose illness as well as allergy and food intolerance, on the basis that every organ dysfunction is accompanied by a specific muscle weakness. It is said to be 'a remarkable and precise diagnostic system' used 'to uncover what your own body "knows" about why you are sick or in pain' and is offered by many osteopaths and chiropractors.[39] '*Painless! Understand your unique needs!*' urges a Bristol AK practitioner.[40] One frequently-used form of this method is for the patient to lie down and raise one arm. The tester first pushes down the arm to determine its muscle resistance. This acts as a baseline. A particular food or other substance is then placed either under the patient's tongue or in a sealed container nearby. Again the tester pushes down the arm. If the arm strength is weaker the second time this is supposed to show the patient to have either a food intolerance or a specific deficiency. The patient may also place the other hand over specific organs to see whether there is any accompanying weakness in the tested arm.

The idea is a mishmash of chiropractic and Traditional Chinese Medicine, with practitioners claiming to be able to evaluate the flow of Qi energy through the body. Magician and sceptic James Randi reports that AK practitioners have claimed 'that while refined sugar can be clearly shown in a very dramatic manner to be a "bad" substance by this method when actual sugar is placed in the hand, the same strength of effect is also brought about by simply having the subject hold a scrap of paper with the word "sugar" or the chemical formula $C_{12}H_{22}O_{11}$ (sucrose) written on it in place of the actual substance. Believers in AK have no problem rationalising the absurdity of such a claim.'[41]

Randi explains that 'the effect observed is entirely due to the expectations of the operator.' The product of suggestion in determining muscle strength and movement without the subject's conscious intention is known as the ideomotor effect and is surprisingly powerful.[42] It was first discovered as long ago as 1833 when the French natural scientist Michel-Eugene Chevreul investigated the claims of dowsers and mediums, describing, in an open letter to physicist André-Marie Ampère, how unconscious and involuntary muscle movements could be initiated by autosuggestion.[43]

You can experience the ideomotor effect yourself by following these instructions taken from the website everything2.com. You will need:

- a 30-35 cm length of cotton thread
- a small weight (1gram or less – a paperclip or small ring will do) tied to one end of the thread
- a sheet of paper, where you have drawn a pattern comprising an equal-armed cross (of about 7 cm) and a somewhat smaller (5-6 cm) circle, with its centre where the arms of the cross intersect.

'Sit at a table and put the paper with its pattern in front of you. Resting your elbow on the tabletop, hold the sewing-thread between your thumb and forefinger so that the pendulum weight (paper clip or ring) hangs straight down, 1-2 cm above the centre of the cross pattern. Try to hold the pendulum as still and immovable as you can.

Now try to imagine how it would feel if the pendulum were to start swinging vertically (in the direction of the vertical arm in the cross pattern). Don't do anything to make it swing, just imagine as intensely as you can the pendulum swinging all by itself. What you will find is that after some 30 seconds to three minutes of intense imagining the pendulum will start to swing. At first these will be small swings, but they will soon grow to wide, bold swings in the vertical direction. It may not necessarily work the first time but in the end it almost certainly will. When you have succeeded with the vertical swings, try to imagine what it would feel like if the pendulum instead started rotating for example clockwise. Again the pendulum reads your mind and will start rotating in the way you imagined. You do absolutely nothing, and still your thoughts somehow make an inanimate object in your hand move in a mentally determined way.'[44]

In Applied Kinesiology, when both the subject and the practitioner are blinded to the test substance, the evocation of the ideomotor effect is no more reliable as a testing method than random chance.[45] [46] Once you've done this experiment it's easy to see how persuasive the ideomotor effect might be.

Could it be magic?

Beyond these social and physical explanations for the attraction of alternative medicine comes the suggestion that our brains appear to be hard-wired to believe in magic. A longer view of human belief reveals evidence that people from widely different cultural and historical periods have always been strongly drawn to what is known as magical thinking.

Magical thinking is a type of non-scientific, superstitious form of causal reasoning to which we are all susceptible. It exists in every society and culture and ranges from the simple superstition – wearing a lucky scarf to a football match – to the highly complex, like a cargo cult in which Pacific islanders believe that their rituals will make a plane or boat laden with desirable goods appear. Benedict Carey in the *New York Times* reminds us that magical thinking is far more common than generally realised and suggests that the brain may have networks that produce an explicit, magical explanation in certain circumstances when it 'leaps to conclusions before logic can be applied'. Magical thinking starts when we are toddlers. Jacqueline Woolley, professor of psychology at the University of Texas, told Carey that by the age of three 'most know the difference between fantasy and reality, though they usually still believe (with adult encouragement) in Father Christmas and the Tooth Fairy'. She went on to note that 'the point at which the culture withdraws support for belief in Santa Claus and the Tooth Fairy is about the same time as it introduces children to prayer'. Carey also describes a research study in which 'young men and women instructed on how to use a voodoo doll suspected that they might have put a curse on a study partner who feigned a headache'.[47]

Magical beliefs and ways of thinking are what fundamentally differentiate CAM from orthodox medicine. Anthropologist Phillips Stevens Jr says such thought patterns are based on principles of cosmology and causality and are timeless and universal. As we have seen, belief in the existence of supernatural forces manifests itself as the notion of 'vitalism' or the 'life force', present throughout alternative medicine. Qi energy, homeopathy's *vis vitalis*, chiropractic's 'innate intelligence', naturopathy's *vis medicatrix naturae* or Ayurveda's 'prana' share many characteristics – not least the fact that they cannot be objectively measured or identified by any means.

'Various other "alternative" and "New Age" beliefs are obviously magical; many are ancient and widespread,' says Stevens. 'Crystals have long been believed to contain concentrated power; coloured crystals have specific healing effects, as certain colours are associated with parts of the body – as they have been in the West for centuries. Colours enhance powers ascribed to candles and other ritual devices.'[48] In the nineteenth century Welsh father and son John and Henry Harries practised in Harley Street and the village of Cwrt-y-cadno in Carmarthenshire. As Fellows of the Royal College of Surgeons, these two doctors offered conventional medical treatment together with what they called 'Maypole healing' in which coloured ribbons were used to encourage patients to visualise and defeat symptoms.[49] Today the 'Miracle of Colour Healing' is still obtainable: Aura Soma® 'is a system that brings you closer to the understanding of yourself'. Developed in 1984 by Vicky Wall, a UK chiropodist who says she developed psychic powers after losing her sight, it 'uses the visual and non-visual energies of colour, it uses the energies of herbs from essential oils and herbal extracts and it uses the energies of crystals and gems'.[50] The means and the marketing may have changed over the years, but the fundamental assumptions and the reasons these therapies appear to 'work' remain the same.

The Golden Bough: A Study in Magic and Religion is a twelve-volume study of mythology and comparative religion written by Sir James George Frazer in the early twentieth century. His work has had a huge influence on contemporary thought and culture and was famously controversial at the time for elucidating the pagan roots of Christianity. Frazer explored the common elements of all religions and, in particular, examined the roots of symbolism and magical thinking or what he called 'sympathetic magic', which he found in all cultures. He described what he called 'the mistaken association of ideas' which manifests itself as two sorts of sympathetic magic, listing hundreds of historical and contemporary examples.

Frazer's observations have striking echoes of the ways of thinking deployed in CAM today. What he terms 'homeopathic magic' is based on the idea that like produces like, with the magic working because things or actions resemble other things or actions. It is invoked in the making of voodoo dolls, the administering of yellow substances to cure jaundice and is exemplified in the case of 'the eminent novelist, Mr Thomas Hardy, who was told that the reason why certain trees in front of his house, near Weymouth, did not thrive, was that he looked at them before breakfast on an empty stomach'. In 'contagious magic' things which have 'once been conjoined must remain ever afterwards, even when quite dissevered from each other, in such a sympathetic relation that whatever is done to the one must similarly affect the other'. Contagious magic can involve the use of charms or amulets, the burying of body parts such as an umbilical cord or placenta, the veneration of saintly relics, or the dipping of one of Ann Callaghan's rocks in a bucket of water. A single contact is all that's necessary. Frazer quotes Pliny's advice that 'if you have wounded a man and are sorry for it, you have only to spit on the hand that gave the wound, and the pain of the sufferer will be instantly alleviated.'[51] Contagious magic can work positively or negatively, such as in homeopathy when you may be warned, with no explanation, rational or otherwise, that you should 'try not to touch your tablets as this may inactivate them'.[52]

Please yourself

In his essay 'The Psychopathology of Fringe Medicine' Karl Sabbagh concedes that alternative medicine often 'works' but says 'when it works it works for none of the reasons given by fringe practitioners themselves'.[53] Sabbagh identifies three reasons why CAM 'works'. First, most of the illnesses it treats are relatively benign, like back pain, tiredness or wind. Secondly, there is a natural variability of all diseases and most get better of their own accord if you wait long enough – it's called regression to the mean. There are few 'quick fix' alternative medicine methods – instead, treat the patient for weeks or months and the odds are that they will recover. Thirdly, many of the ailments that CAM treats have subjective or psycho-somatic aspects that are responsive to a placebo.

The so-called placebo response is when symptoms are altered for the better by an otherwise inert treatment, a 'dummy' pill for example. It is often cited as the explanation for the apparent success and popularity of many CAM therapies. But in any discussion of placebo we need to make a distinction between the effect on the *outcome* of disease and the effect on the *experience* of illness. As Peter Skrabenek and James McCormick write in *Follies and Fallacies in Medicine*, 'disease may or may not be accompanied by illness. Many diseases, including some that are potentially serious, are often symptomless. On the other hand, feeling unwell is not always the result of disease. Placebos have no effect on the progress or outcome of disease, but they may exert a powerful effect upon the subjective phenomena of illness, pain, discomfort, and distress'.[54] In other words, placebos may make us feel better, but feeling better does not necessarily constitute a cure.

'The placebo effect is part of the human potential to react positively to a healer,' Michael Jospe, a professor of psychology, told the magazine *FDA Consumer*. A placebo reduces distress even when there is no material reason why it should. 'When you put a Band-Aid on a child and it has stars or comics on it, it can actually make the kid feel better by its soothing effect,' says Jospe.[55] Sometimes we feel better simply because we expect to feel better – this is known as subject expectancy. Subject expectancy can bring about an increase in the production of pain-relieving chemicals called endorphins and enkephalins in the brain, as research at the University of Michigan and elsewhere has shown.[56] Neuroimaging has also revealed measurable biological changes in the brain after placebos have been given.[57]

Debate continues about the placebo response – how effective it really is as well as what psychobiological effects it might prompt and whether it is ethically acceptable to exploit it. It has been a subject for active research since the 1950s when Henry K Beecher published a research paper 'The Powerful Placebo' indicating that an average of 32 per cent of patients responded to a placebo.[58] However, recent research comparing the use of placebo with no treatment in a total of 166 randomised controlled trials has found little evidence of any powerful clinical effect and no evidence that a placebo can actually alter the natural course of an illness.[59] The placebo can only be said to be genuinely effective if it produces a response greater than doing nothing.

One theory for how the placebo response may work is that it suppresses the complex biological phenomenon known as the acute phase response. This includes the four classic signs of inflammation: swelling, redness, heat and pain. 'Besides these physical changes, there are also important psychological ones, including lethargy, apathy, loss of appetite and increased sensitivity to pain – a suite of symptoms that are collectively known as sickness behaviour,' says psychologist Dylan Evans. He suspects that placebo does not work so well in illnesses without an acute phase response.[60]

There is evidence that in some cases the placebo response may be as effective as surgery, as one American study found when 180 patients with osteoarthritis of the knee were randomly assigned to receive genuine surgery or placebo surgery. Patients in the placebo group received skin incisions and underwent a simulated procedure. Afterwards there was no clinically meaningful difference between the two groups – the reduction in pain and improvement in knee function were more or less the same.[61] But rather than there being an immutable 30 per cent placebo response, it appears to vary according to illness. People with pain conditions such as diabetic neuropathy have been shown to be more responsive to placebo than those suffering the pain that can occur after a bout of shingles.[62] Various mechanisms have been suggested: the relief of anxiety, expectation, transference, 'meaning effects' and conditioning.

The effects of a dummy drug, however, aren't always entirely positive. The nocebo (Latin for 'I will harm') is what's been called the placebo's evil twin.[63] It is an inert remedy which induces unpleasant effects, possibly caused by similar mechanisms to the placebo response – anxiety and expectation, for example. Ethical concerns have limited research in this area, but there have been suggestions that negative nocebo effects may be more frequently evoked than we realise. A recent review found that patient expectation of possible unpleasant or harmful side-effects of a drug or treatment may influence a treatment's outcome.[64]

It gets even more interesting when we discover that – for the same ailment – some types of placebo are more effective than others. A study from Harvard Medical School compared the use of sham acupuncture with a dummy pill for persistent arm pain caused by Repetitive Strain Injury (RSI). Overall, the sham acupuncture device had greater effects than the placebo pill, leading the researchers to conclude that 'placebo effects seem to be malleable and depend on the behaviours embedded in medical rituals'.[65] In short, these rituals are the adult version of the cartoon on the child's sticking plaster. Humans are highly suggestible – it is already known, for example, that the colour of tablets can determine our reactions. Red, yellow or orange placebo tablets have been found to be more stimulating than blue or green ones, which are more calming.[66]

But one element is essential, regardless of the type of placebo used. Professor Edzard Ernst explains that 'what seems important for maximising placebo effects is to administer treatments with empathy, sympathy, conviction etc . . . a good therapeutic relationship may not be the only precondition for generating a sizeable placebo response, but it helps'.[67] Alternative practitioners themselves admit that the placebo is a good way to 'catalyse the process of healing' and certainly the therapeutic relationship engendered by the first altmed appointment, lasting as long as an hour, must significantly stimulate a placebo response that can be reawakened in subsequent, shorter, meetings.[68] Homeopathic remedies in particular 'are personalised and contain a huge quantity of caring – an unmeasurable essence contained in a deeply personal gift, over which providers have spent much effort' writes infectious diseases consultant and critic of homeopathy Dr Philip Welsby. 'Who would not be impressed and affected?'[69]

In a provocative development Alun Anderson, a former editor of *New Scientist*, has suggested that 'alternative medicine could focus on finding ways to make the [placebo] effect more powerful instead of pretending to offer the same benefits as conventional medicine'. Controversially, he also suggests the mainstream should be more willing to take on the ways of CAM. 'Scientists who debunk alternative medical treatments should think again about whether it is ethical to destroy people's belief . . . further in the future, we will learn how both active drugs and belief and rituals work inside the brain. The distinctions between mind and body will break down and we'll have a new, improved, unified science of medicine that includes a healthy dose of snake oil.'[70] Psychiatrist David Spiegel is all in favour: 'most placebos are relatively harmless,' he says. 'Modern medicine involves treatments, such as surgery, chemotherapy, and bone marrow transplantation, that are effective but also toxic. Many patients may choose integrative medicine as a kinder and gentler treatment that harnesses rather than eschews the placebo effect and engages them as participants in their care, especially in the treatment of chronic problems such as anxiety and pain that are often not well managed in medicine'.[71]

Medical historian David Wootton also thinks that homeopathy deserves a place in the mainstream. 'What homeopathy can do for you is thus a good indication of what a placebo can do for you, and, on the definition I am proposing homeopathy does neither good nor harm, though it is perfectly reasonable that it should be available (as it increasingly is) on the National Health Service, since it performs much better than no treatment at all.'[72] The proposal appears to be that we should be encouraged to accept anything a doctor gives us, regardless of whether or not it is a real drug or that the prescription entails deception. In any case, CAM users may not care. One devoted user of alternative medicine tells me she doesn't care if the homeopathic treatment she believes cured her daughter's tonsillitis was a placebo. All that matters to her is that the remedy appeared to 'work'.

GP Margaret McCartney, in her article 'Reclaiming the placebo', says she thinks there is considerable potential benefit: 'misunderstood and often derided, placebos need active consideration, not least because new medicines often offer only marginal gains in effectiveness . . . in clinical practice, the placebo effect is one of the most useful, consistent, cheap, side-effect-free and convincing things that medicine has to offer'.[73] But writing in the *British Medical Journal*, Zelda Di Blasi and David Reilly disagree. They say 'placebos may seem like a less toxic solution than pharmacological treatments for functional or chronic conditions but this carries side effects – disrupt trust, and outcome is disturbed . . . The issue should not be about prescribing placebos but rather about the need to increase our general knowledge around healing mechanisms to harness directly what placebo harnesses indirectly, in an ethical and practical manner, encouraging a sense of trust and partnership between the public and healthcare specialists'.[74] This aim of transparency and the honest sharing of knowledge has to be preferable to enduring either the falsehoods of alternative medicine or the anachronistic paternalism of an old-fashioned doctoring, whose authority was founded on the withholding of information and the benign mendacity of the prescription of a placebo 'tonic' a day to keep patients at bay.

Of course, placebo response will never just be confined to users of alternative medicine. Everyone has the capacity to respond positively to a personally effective physician. A confident and empathic mainstream practitioner will evoke her patients' placebo response alongside all the other material benefits of conventional medical treatment, whatever form it takes. Dylan Evans may argue that 'the high levels of consumer satisfaction among users of alternative medicine suggest that alternative therapists are better at evoking confidence and trust in their patients, and so better able to mobilise the placebo response, than are conventional doctors'.[75] But surely CAM practitioners learn to do it better because, in truth, it is all they have.

Chapter 10

The Healthy Sceptic

There are two definitions of the word 'sucker' I had in mind when I thought about the title of this book. *Sucker*: one who lives at the expense of others, and *sucker*: a gullible or easily deceived person. The growth of alternative medicine in the modern era has depended on the existence of both kinds of sucker.

And when I suggest that alternative medicine makes fools of us all, I am thinking about not only those who practise alternative medicine and those who use it, but also those of us who simply let it carry on unchallenged. There is a pervasive tolerance of complementary and alternative medicine which allows it to flourish. I believe this indulgent laissez faire approach is no longer acceptable and needs to change. The growth of CAM matters – it trades in false hope, it is bad for our health, it threatens our intellectual culture, it wastes public money and it undermines some of our most important and valued institutions. Those who promote it and the junk science that all too often accompanies it, are, to quote Raymond Tallis, 'enemies of mankind's best hope for a better future. In short, enemies of mankind'.[1]

Making money

There is no doubt that the profit motive is at the heart of both the private health sector and the pharmaceutical industry, but it is by no means their sole motivation. And their activities have ultimately to be checked by many processes of accountability which expose risks as well as benefits. In the world of CAM there are no such checks. The desire by the practitioner to maximise profit can have an unlimited impact on their practice and it's simply nobody else's business. Dan Hurley's investigative book *Natural Causes*, for example, has exposed how the American herbal and vitamin supplements industry escapes regulation while making millions of dollars selling products which are neither safe nor effective. Much of what he reveals is also relevant to the industry in the UK and Europe.[2]

The truth is that no one knows how much money is being spent on CAM in the UK. Estimates of how much 'out of pocket' personal cash is spent vary from £164 to £648 per user per annum, but these figures are several years old and are likely to underestimate current spending.[3] [4] Boots has seen sales of herbs, for example, more than double in four years, from £17 million in 2002 to £47 million in 2006.[5] The Smallwood report on the role of complementary medicine in the NHS suggested that up to £4.5 billion was spent in 2005 on CAM usage in the UK, with around 10 per cent of that sum − about £450 million − being spent by the NHS.[6]

But this figure is little more than conjecture, as the Department of Health itself has no idea how much it spends on alternative therapies in the NHS. CAM spending is not centrally audited and local health bodies − which include hospitals, primary care trusts and GPs − can buy as much or as little of it as they choose, without reporting this spending to central government. Phil Willis MP, chair of the House of Commons Science and Technology select committee finds it 'very strange that this information is not available to the public already' and thinks 'it is clearly critical to know how much we are spending on this area if the Government is to make balanced decisions about alternative medicine that are properly informed by evidence'. He is supported by fellow committee member Dr Evan Harris MP, who says 'it is not good enough for the Government to say that it does not know how taxpayers' money is being spent on therapies that have not been shown to work or have been shown not to work'.[7]

According to the then health minister Caroline Flint 'it is the responsibility of local NHS organisations to commission healthcare packages for NHS patients, be it complementary or orthodox'.[8] In practice this means NHS hospitals can decide to hire practitioners like Graham King, who performs reiki healing two days a week at the children's cancer unit of the Middlesex Hospital. He's there because he's popular with parents, staff and children who say they find reiki relaxing and calming; King, for his part, says he is channelling cosmic energy into the children's bodies. No one begrudges cancer patients access to any form of relaxation as part of their care, but it is surely unacceptable when relaxation is delivered in the form of reiki hocus pocus which deceives patients and raises false hope about clinical improvements to serious disease. One mother told Julia Stuart of the *Independent on Sunday* that she believed her son Martin would be cured of leukaemia 'both through reiki and the medication'.[9] Fortunately her son's form of leukaemia has a good cure rate.

News has also emerged recently that University College Hospital in London is spending £80,000 per year on a team of ten spiritual healers who are despatched to patients' bedsides in the hope that their ministrations could raise the white blood cell count after chemotherapy.[10] Here is how the posts were advertised, offering us a disturbing vision of the 'integrated' future for Britain's health service:

University College London Hospital NHS Healer Cancer / Haematology unit

Position Title: Healer / Complementary Therapist
Hospital Site: University College Hospital London Euston
Pay Band/Grade Band 5
Minimum Salary (P.A.Inc): £22,999.00
Maximum Salary (P.A.Inc): £29,764.00
Contract Type: Fixed Term
Description: 12 month fixed term contract 28 hours per week

We are seeking a motivated practitioner to join our complementary therapy team which offers aromatherapy, counselling, healing, massage and reflexology to our patients. The successful candidate will hold a Reiki Master/Teacher Certificate via the Usui system with two years post certificate patient experience, ideally within a hospital setting, or a full National Federation of Spiritual Healers

Certificate with two years post certificate patient experience, ideally within a hospital setting.

The team works in the largest Haematology Unit in the UK, helping to provide holistic care to patients with diseases such as leukaemia and lymphoma including patients undergoing intensive bone marrow transplants. Expansion of our service means that we are now also working with oncology patients within cancer services.[11]

By funding alternative practitioners in this way the NHS confers a seal of approval and support to quackery, misleading patients into perceiving some genuine influence on the outcome of a disease. It amounts to an active endorsement of alternative medicine, thereby encouraging the public appetite for yet more worthless therapies. Nor is it the case that the health service has funds to squander on this kind of foolishness – one recent inspection found that half of England's hospital kitchens fail to meet basic hygiene levels and another survey confirmed that many elderly hospital patients in the UK become malnourished because no one can spare the time to help them eat.[12] [13]

In Northern Ireland, meanwhile, £200,000 of government money has been paid to a private company, GetwellUK, 'to pilot services integrating complementary medicine into existing primary care services in Northern Ireland'. GPs in two areas will be funded to refer patients for therapies like acupuncture, homeopathy and massage. Former NI Secretary of State Peter Hain is 'delighted that Northern Ireland is leading the way in integrating complementary and alternative therapies into the National Health Service'. Peter Hain believes his son's eczema was cured with homeopathy and dietary changes: 'I am certain, as a user of complementary medicine myself, that this has the potential to improve health substantially.' In his 2005 speech to the Prince's Foundation for Integrated Health, Peter Hain underlined the role of the market in the planning of health provision and made the dangerous suggestion that patient power should dictate NHS provision of CAM, irrespective of evidence. 'Ultimately the future will be driven by the patients. And patients want real choice: the best of all worlds, combining the benefits of both complementary and conventional treatments, free on the NHS.'[14]

In Scotland, 60 per cent of GP practices now prescribe homeo-pathic or herbal remedies. As ever, the majority of the patients prescribed homeopathy are female, with a median age of 48.[15] Acupuncture is also well established, according to a media release from NHS Highland who report that 'an increasing number of NHS Highland patients are enjoying blissful relief of pain thanks to an ancient Chinese medicine' administered by thirty NHS High-land physiotherapists.[16]

The Prince Charles' Foundation for Integrated Health styles itself as 'a powerful force for creating a true health service'. Its aim is to help 'people, practitioners and communities to create an integrated approach to health and wellbeing', to which end it has been given £900,000 by the UK Department of Health. The foundation claims that 'integrated health acknowledges that humans are complex self-healing organisms' and points to evidence that three-quarters of the British public support the use of comple-mentary therapies on the NHS. The approach the foundation favours is exemplified by its Awards for Good Practice: past winners include the Holistic Midwifery Service at the George Eliot Hospital NHS Trust which offers aromatherapy and yoga and the Blackthorn Medical Centre and Trust in Barming, Kent, a GP practice which uses Rudolf Steiner-inspired Anthroposophy which involves art, music and movement for patients with chronic illness.[17] Everything in the garden sounds so lovely that it seems almost churlish to mention that the best that can be said for these practices is that they might occasionally enhance the placebo response. And once again, while they can offer relaxation and reassurance at a difficult or challenging time, they should not be given spurious endorsement as effective therapies.

It is arguable that the 'integration' of complementary medicine would not have been so successful in the UK were it not for the personal predilections of Prince Charles and his mother. The Prince of Wales has had a number of meetings with officials from the Medicine and Healthcare products Regulatory Authority, though allegations that he has tried to influence the authority and promote the FIH agenda behind the scenes are denied by his principal private secretary, Sir Michael Peat.[18] [19] Maybe it takes an American to tell it like it is.

The editor of the *Federation of American Societies for Experimental Biology Journal* Gerard Weissman is certainly not impressed by such reports of royal support for CAM: 'one notes that when Prince Charles and other fans of unproven or disproven medical practices use terms such as "integrated therapy" or "alternative medicine" they're following the lead of creationists who hide under the term "intelligent design" – these are all convenient slogans that permit the credulous to con the gullible.'[20]

The Royal family are the most famous users of the Royal London Homeopathic Hospital in London and the Queen is its patron. And whilst the National Health Service may be unwilling to pay for expensive cancer drugs, University College London Hospitals Trust had no qualms about spending £20 million on the refurbishment of the Royal London Homeopathic Hospital which, amongst other things, dispenses sugar pills to an eager public. No matter, as Professor Michael Baum put it, 'if that sum of money was spent on making available Herceptin and aromatase inhibitors, then it could be a saving in my own health district of six hundred lives a year.'[21] It costs the trust at least a further £3 million every year to run and further funding comes from those Primary Care Trusts who refer patients.

Alarmed perhaps by the rate of acceptance and integration of CAM into the mainstream, some leading British doctors and academics are at last encouraging the NHS to turn its back on alternative medicine, and on homeopathy in particular. In 2006 a thirteen-strong group of scientists, including a Nobel prize winner and six fellows of the Royal Society, wrote an open letter to all NHS Trusts arguing that unproven and disproved treatments should no longer be publicly funded and asking them to review their practices. 'At a time when the NHS is under intense pressure, patients, the public and the NHS are best served by using the available funds for treatments that are based on solid evidence,' they wrote.[22] They sent a further letter along the same lines a year later.[23] Many trusts responded by looking at outpatient provision and by the middle of 2007 more than half the Primary Care Trusts in England were restricting referrals to the UK's homeopathic hospitals or refusing to pay for them altogether.[24]

Supporters of the Royal London Homeopathic Hospital are fearful that without direct funding it may have to close. In a letter seeking support from sympathisers, its clinical director Dr Peter Fisher described what he called 'an orchestrated campaign' against the hospital and employed the by now familiar 'people like it' defence: 'our unique approach enables us to treat successfully chronic complex conditions and our patients have a very high level of satisfaction with our services'.[25] He has reiterated this view elsewhere, together with a more than homeopathic dose of ageism: 'homeopathy is enigmatic: remarkably popular, widespread and persistent, despite the scepticism of retired professors of biomedical background,' he wrote.[26]

The RLHH should be required by the Department of Health to produce more convincing reasons for why it should be supported by the taxpayer, regardless of the royal patronage it enjoys.

Professor Michael Baum has, however, thought of a much better purpose for the Royal London Homeopathic Hospital building. He says the NHS should 'stop peddling placebos and turn the hospital into a centre for evidence-based, supportive care for people with life-threatening or terminal illnesses. A centre with psychologists, masseurs, counsellors, art and music therapists. Unlike homeopathy, these therapies have been critically evaluated: they are proven to enhance well-being. And add a research centre so we can further this area of healthcare. This will make a real difference to people's quality of life.'[27]

Although other health authorities may not be spending so much, the costs nevertheless have an impact on general health provision when resources are already stretched to the limit. West Kent Primary Care Trust says it spends about £200,000 per year on homeopathy. This funds about 2,800 homeopathy outpatient appointments for around 350 people, most of which take place at Tunbridge Wells Homeopathic Hospital, amounting to a spend of £250 per person. Most homeopathic patients receive other specialist services for their condition so the PCT is 'paying twice' for their care. To put these figures in context 'the total amount spent on homeopathy amounts to about 1,500 appointments with pain specialists, 2,300 appointments with dermatologists, or 1,300 appointments with arthritis specialists'. The PCT consulted local people about whether the NHS should pay for homeopathy treatment which led to a decision to stop funding it from 2008.[28]

But other moves toward integration proceed apace. More than 5,000 alternative therapists are listed on the NHS Directory of Complementary and Alternative Practitioners, a volume made available to all GPs, Primary Care Trusts, practice nurses and health visitors. Patients can also use the directory to find local practitioners who are willing to take NHS referrals who can then be contracted and paid out of the NHS budget. It costs therapists just £50 per year to get on the list – they only have to be 'in good standing in one of the leading CAM organisations' and to have the requisite insurance. Directory members receive a certificate for public display that describes them as an 'NHS Trusts Association Complementary Practitioner' for use on websites, publicity leaflets, letterheads and business cards.

And whilst members are warned not to claim they 'are "registered" or "recognised by" the NHS', with this apparent imprimatur it's not as if they need to. The directory's own description of what is available includes everything from craniosacral therapy ('as a general rule, three to ten sessions will make a significant impact on most people's health') to Chinese Herbal Medicine ('If the flow of Qi in the meridians becomes blocked or there is an inadequate supply of Qi, then the body goes out of balance').[29] The NHSTA, which represents all the primary care and hospital trusts in the UK, also publishes *Complementary Alternatives* magazine, 'the place to find some of the best information on therapies, products and ways of life from the world of alternative healthcare'.[30]

The university of lies
NHS provision is, however, just one of the ways in which we are being suckered by CAM. Thousands of students now study alternative medicine in British universities and colleges. In 2008, according to the university application service UCAS, there are sixty-eight CAM courses, forty-nine of them at degree level. These courses can also be a means of obtaining a licence to practise and result in a Bachelor of Science honours degree. There's one in

Aromatherapy at Anglia Ruskin University and another in Ayurvedic Medicine at Middlesex. Complementary Therapies (Stress Management) is an option at Greenwich and the University of Derby offers a BA (Hons) in Accounting and the Healing Arts.

Students at these institutions are invariably taught by alternative practitioners, not critical thinkers who might examine objectively the evidence of altmed claims. That universities can decide for themselves what courses they offer, based on nothing but whether they think they can fill the places, poses no problem according to the umbrella organisation for all British universities, Universities UK. They ruled that it was up to individual institutions to decide how they describe particular courses.[31] What is more, external examiners, employed to maintain the standard of degrees, are usually representatives of the altmed establishment. UCL professor of pharmacology David Colquhoun has investigated these 'science degrees without the science' and argues convincingly that whilst courses in subjects such as golf course management and surf studies may have been widely derided as 'Mickey Mouse' at least 'degrees in things such as golf-course management are honest. They do what it says on the label. That is quite different from awarding BSc degrees in subjects that are not science at all, in fact positively anti-science. In my view they are plain dishonest'.[32]

The number of complementary medicine students has grown in the last couple of years to reach around 7,000 in 2006, with complementary medicine enjoying a bigger surge in popularity than any other university subject. In 2006 acceptances increased by 36.5 per cent, no surprise when one finds that it is so easy to secure a place.[33] The minimum academic requirements for a BSc in Homeopathic Medicine at the University of Central Lancashire, for example, amount to only one B and one C grade at A level and just three grade Cs in English, Maths and Science at GCSE. In contrast, those who want to do a course in scientific medicine at the University of Newcastle must have three As at A level and these must include Biology and Chemistry.[34]

Likewise only a B and a C in any subject are required for the Complementary and Health Sciences BSc degree at the University of Salford, who say CAM 'forms a rapidly increasing proportion of how health care in the UK is delivered and is an important area of academic study'. At Salford you can become professionally accredited in Reflexology and Aromatherapy at the same time as doing your degree. 'If you complete all of the homeopathy modules in the degree, you will be able to apply for entry into year two of the four-year part-time homeopathy professional training programme at the North West College of Homeopathy in Manchester'.[35] Perhaps more alarming for the rest of us is the information that 'you will also receive a certificate in Homeopathic First Aid'. In an address to prospective students at the University of Westminster, Head of Complementary Therapies Brian Isbell was relaxed about the lack of research validation for CAM, promising that 'evidence will come once it's within mainstream provision'.[36] At present these degree courses cost UK students £3,000 every year. A rough estimate, based on the number of students and the likely cost of these courses suggests that they are subsidised by the government to the tune of between £13m and £17m per year.[37]

These supposedly academic institutions are reminiscent of Jonathan Swift's 'Grand Academy of Lagado', at which Gulliver encounters a professor engaged in 'a project for extracting Sun-Beams out of Cucumbers, which were to be put into Vials hermetically sealed, and let out to warm the Air in raw inclement Summers'.[38] Here, for the record, are fifty government-funded UK universities and colleges offering CAM degree courses for 2008, as listed by UCAS: Anglo European College of Chiropractic; Anglia Ruskin University; University of Bedfordshire; Birmingham College of Food, Tourism and Creative Studies (college accredited by the University of Birmingham); Blackpool and The Fylde College, an associate college of Lancaster; University of Brighton; City of Bristol College; British College of Osteopathic Medicine; British School of Osteopathy; University of Central Lancashire; Cornwall College; University of Derby; East Lancashire Institute of Higher Education at Blackburn College; University of East London; Edge Hill University; Farnborough College of Technology; University of Glamorgan; University of Greenwich; Harper Adams University College; University of Huddersfield; UHI Millennium Institute; Leeds Metropolitan University; University of Lincoln; Middlesex University; Napier University Edinburgh; Newcastle College; New College Durham; The North East Surrey

College of Technology; The North East Wales Institute of Higher Education; Northumbria University; Oxford Brookes University; Pembrokeshire College; University of Plymouth; Riverside College Halton; University of Salford; Southampton Solent University; St Helens College; Stockport College; Stratford-upon-Avon College; University Campus Suffolk; Swansea College; Thames Valley University; Truro College; Tyne Metropolitan College; University of Wales Institute, Cardiff; University of Westminster; Wigan & Leigh College; University of Wolverhampton; University of Worcester.

As the chiropractic school pioneer B J Palmer must have understood a century ago, training alternative practitioners provides a guaranteed income only for those who provide the training. Evidence is beginning to accumulate that the yearly crop of newly qualified naturopaths, herbalists and reflexologists are encountering difficulties when finding employment in an already crowded marketplace. The *Guardian* quoted one osteopathy graduate as saying that 'in London, the competition for jobs and patients is huge. Lots of friends that I graduated with are only doing one or two days' osteopathy per week. To make a living, they have to spend the rest of their time working in bars or restaurants'. Even those who manage to land a rookie slot as an 'associate' in an established osteopathic practice for example will need to advertise themselves. 'I expect my associates to help write letters of introduction to GPs, midwives and other practitioners in the area, to help contribute to articles in specialist magazines or the local paper; and to network with the local community,' says North London osteopath and clinic founder Gavin Burt.[39]

Science is golden

The Enlightenment of the eighteenth century helped us move away from an approach to our bodies and our health governed by nothing more than magic and faith. Doctors were no longer able to make empty assertions and quacks could not invent cures with impunity. At the heart of the Enlightenment was science, but today we seem to have become confused about what science is. Science does not necessarily mean scientists or technology or a fixed body of facts under the rigid subheadings of chemistry, physics and biology; what it fundamentally stands for is an approach to knowledge. Science offers us a way of understanding and making sense of the world around us. And the scientific method amounts to a general framework which can entail the use of different research techniques depending on what is being studied. If you reject science you reject rationality itself. We need to understand and value science if we are successfully to counter the attempts by CAM to replace it with superstition and chicanery.

Using the model of hypothesis, experiment, analysis of evidence, peer review and publication of results, science provides us with the demonstrably best way of establishing how to tackle disease and illness. It is an essentially humble practice: it tells us that we don't know anything without evidence and that the results of our investigations as well as our practices must always be open to scrutiny and challenge. Physics professor Rory Coker deserves quoting at length on this:

> Science relies on – and insists – on self-questioning, testing and analytical thinking that make it hard to fool yourself and to avoid facing facts. Pseudoscience, on the other hand, preserves the ancient, natural, irrational, unobjective modes of thought that are hundreds of thousands of years older than science – thought processes that have given rise to superstitions and other fanciful and mistaken ideas about man and nature – from voodoo to racism; from the flat earth to the house-shaped universe with God in the attic, Satan in the cellar and man on the ground floor; from doing rain dances and to torturing and brutalising the mentally ill to drive out the demons that possess them.[40]

Science nevertheless has to be open minded and prepared to change its views when it meets new evidence that contradicts earlier findings. Its most fundamental assumption remains that all suggestions must be open to question. As Richard Feynman wrote in his famous essay decrying what he called 'cargo cult science', 'if you're doing an experiment, you should report everything that you think might make it invalid – not only what you think is right about it: other causes that could possibly explain your results; and things you thought of that you've eliminated by some other experiment, and how they worked – to make sure the other fellow can tell if they have been eliminated'.[41] Unfortunately, as Raymond Tallis explains, alternative practitioners make the most of this explicit fallibility. 'The honesty of science about its limitations, its often reported failures, provides an ideal environment for anti-science, junk science, to flourish, which in part depends upon exploiting the problems and failures of science. Even so, junk science should have nothing going for it; it should be laughed to scorn by even a half-educated public.'[42]

To base our approach to health and medicine on anything other than the scientific method is both foolish and dangerous. Standards of evidence cannot be adjusted – why should there be one standard for testing a homeopathic remedy and another for testing a cancer drug? There is no reason why they should not both be subject to the same values of both methodology and ethics. As Richard Dawkins wrote in his foreword to John Diamond's book *Snake Oil and Other Preoccupations*, 'either it is true that a medicine works or it isn't – it cannot be false in the ordinary sense but true in some "alternative" sense'.[43] Or to put it another way, as expressed in an editorial in the *Journal of the American Medical Association*: 'There is no alternative medicine. There is only scientifically proven, evidence-based medicine supported by solid data or unproven medicine, for which scientific evidence is lacking. Whether a therapeutic practice is "Eastern" or "Western," is unconventional or mainstream, or involves mind-body techniques or molecular genetics is largely irrelevant except for historical purposes and cultural interest.'[44] Extraordinary claims – like those made by homeopathy – require extraordinary evidence. As biochemist Thomas J Wheeler says, 'we are entitled to ask for very strong evidence before we accept ideas that are contradicted by a large amount of prior evidence.'[45]

It is also true of course that while modern, evidence-based medicine cures some of the people some of the time, it is also far from infallible. Doctors and pharmaceutical companies sometimes use approaches that carry more risks than benefits – selling arthritis drugs which can cause heart problems, or over-prescribing the antibiotics which have facilitated the development of so-called 'superbugs' in Britain's hospitals. Drugs can have terrible unforeseen consequences, such as thalidomide. Medical science has been accused of being slow to accept and act on discoveries, like the isolation of the organism *Helicobacter pylori* as a cause of stomach ulcers. Though recognition of *H. pylori* may have seemed to take too long at the time, the fact is that it took only twelve years to establish a complete and worldwide change of practice in ulcer treatment, after which its discoverers won a Nobel prize. It is worth remembering that the proponents of acupuncture, herbalism and the like believe one of alternative medicine's chief virtues to be that it hasn't changed in three thousand.

The nature of evidence is that it can be of variable quality, which means it should be accorded variable weight. The weakest sort of evidence is that particularly favoured by the champions of CAM, the individual account, also known as anecdotal evidence. Your next-door neighbour may tell you that her headache goes away when she wears a red hat, but that does not mean you can extrapolate from her report to say that red hats are a headache cure, however convincing she sounds when she says 'it works for me'. At the other end of the reliability scale is what's usually described as the 'gold standard' in research terms, the double-blinded Randomised Controlled Trial (RCT) whose aim is to produce a research framework that can remove any bias which could skew the results.

Science goes to elaborate lengths to ensure it evaluates evidence with rigour and objectivity. Subjects in an RCT are randomly split into groups, usually by computer, the aim being to produce a treatment group and a control group. Each subject and group should be as much like each other as possible so as to avoid introducing other factors that could influence the results, known as 'confounding factors' or 'lurking variables'. You can then investigate the effects of a test substance – a red hat, say – by comparing what happens to the 'experimental group' who wear one when they have a headache, with the 'control group' which doesn't.

Some research studies may use several groups – the first might be using one particular therapy, the second a placebo and the third receiving no treatment. Or one group may be taking an old treatment for a condition while the other takes a new one. Double blinding means that neither the researcher nor the subjects know the composition of the groups or who is having which treatment, so that their beliefs and expectations cannot affect the results. The trial is therefore randomised, controlled and 'double blinded'. If the researcher knows which are the active treatments but the subject does not, this is 'single blinding'. Even more useful conclusions can be drawn by collecting and reviewing a series of RCTs, offering an opportunity to look at the effects of a particular intervention on perhaps thousands of subjects. This is called a systematic review or meta-analysis.

Levels of evidence

A judgement about the quality of evidence can be informed by grading it along the lines of the list below:

1. Systematic review of multiple well-designed, randomised controlled trials (RCT)
2. At least one RCT
3. At least one well-designed intervention
4. At least one well-designed observational study
5. Expert opinion, including users and carers[46]

As you see, the positive testimonials that usually constitute the only evidence offered by alternative practitioners (aka 'success stories') simply do not figure. Such accounts represent the poorest kind of evidence, of even less value than the tale your be-hatted neighbour tells you over the garden wall. There is no way to ascertain the veracity of testimonials and certainly no way of hearing from those whom the treatment has failed.

The systematic review may provide the most weighty evidence, but that doesn't mean that level 5, which includes the experiences of users and carers, is meaningless or should be disregarded. And even though it may represent the gold standard, the double-blinded RCT is not the only means of securing useful information. Sometimes, for example, blinding is simply not possible. When John Diamond agreed to participate in a trial to see how best to administer chemotherapy for his type of cancer – a dose once every three weeks or continuously via a Hickman line into the chest – it 'could hardly be either single or double blind: even if I hadn't noticed a yard of plastic tubing sticking out of my chest, my oncologist almost certainly would have.' But this research involved other objective methods of measuring what was going on, such as measuring and comparing any change in size of the research group's tumours.[47]

Tricia Greenhalgh, professor of primary health care at University College London, describes evidence-based medicine as 'the use of mathematical estimates of the risk of benefit and harm, derived from high quality research on population samples, to inform clinical decision making in the diagnosis, investigation or management of individual patients.' Her book *How to Read a Paper: The Basics of Evidence-Based Medicine* as well as *Bandolier's Little Book of Making Sense of the Medical Evidence* are both illuminating and detailed expositions of scientific ways to find out and evaluate medical research.[48] [49] Whilst they are aimed primarily at the health professional, both books offer the layperson useful explanations of modern research principles and techniques. It is because scientific method underpins the work of conventional medicine that medicine can be called to account; the proper gathering and analysis of evidence can support a particular therapy but it can equally demonstrate its risks and failures. From this philosophy we have built structures of accountability to support scientific medicine: the National Institute for Health and Clinical Evidence, the Food and Drugs Administration and so on. This philosophy and these structures are there to protect both doctors and patients and need to be strengthened, not undermined by either magical thinking or market forces.

'Most complementary therapists are aficionados, but when you are a researcher your enthusiasm should be first and foremost for good science,' writes Professor Edzard Ernst, who has arguably conducted more scientific research in the area of alternative medicine than anyone else in the world.[50] But alternative practitioners often refuse to accept the results of scientific investigation and RCT-based research. Kathy Sutton's recent report on complementary medicine, funded by GetwellUK, recommended much greater access to CAM therapies on the NHS.[51] She argues that 'because complementary health care has largely been seen through a narrow prism of health policy, the test for comprehensive public provision has been based on clinical tests of efficacy as measured by randomised controlled trials. Yet complementary health care consists of two key elements – clinical and social care – concerned with both the alleviation of disease . . . and the promotion of healthy lifestyles . . . A fuller evaluation of complementary health care should take into consideration both aspects of care as well as issues of public health'.[52] Anything, it seems, is preferable to that troublesome 'narrow prism' that consistently illuminates negative results.

Others argue that the evidence-based approach is just not sophisticated enough to cope with the complexities of complementary and alternative medicine. The argument is difficult to follow, but appears to amount to advising us that the benefits of CAM are so great as to be immeasurable: 'Evidence Based Medicine misses emergent properties of complex systems when those system components lose their power if separated into parts. Healing approaches and many complex integrative systems of medicine present exactly such complex systems. Healing is defined as the process of recovery, repair and reintegration that persons and biological systems continually invoke to establish and maintain homeostasis and function. These processes are the most powerful force we have for recovery from illness and the maintenance of well-being, and so the most important for clinical practice'.[53] It is hard to see how researchers could so easily miss something as powerful and significant.

CAM supporters also argue that RCTs cannot yield useful information in the highly personalised setting of, say, a homeopathic or Traditional Chinese Medicine consultation. The argument usually runs along these lines, as expressed by the British Homeopathic Association: 'in research terms, it is not always appropriate to make a "like for like" comparison between homeopathy and conventional medicine. Comparisons of homeopathic medicines with placebo are not always appropriate or wise. Homeopathy usually takes an individualised approach to treatment, and two patients with the same diagnosis can be treated with very different homeopathic medicines depending on the presentation of symptoms'. Not surprisingly, the British Homeopathic Association favours those so-called 'outcome studies' in which doctors ask their patients whether they feel better after treatment and most of the patients say yes.[54] It is striking, to say the least, that the organisation charged with promoting the interests of one of the most popular CAM treatments chooses to rely on this entirely subjective approach.

Lastly there are the 'true believers', whose confidence in the efficacy of their chosen method is so strong it is unaffected by evidence to the contrary. Psychologist Ray Hyman describes his involvement in a test of Applied Kinesiology, in which a team of chiropractors sought to demonstrate how they could tell the difference between what they saw as a 'good' sugar (fructose) and a 'bad' one (glucose). First they performed Applied Kinesiology in the way I described earlier, with volunteers who lay down and raised an arm while the practitioner placed a solution of one of the sugars on their tongues. The practitioners believe that this experiment would show that the body recognised glucose as a 'bad' sugar, and sure enough, the participants' arms were weaker after they were given it. Conversely, when they tasted the 'good' fructose, their arms were more resistant.

The experiment was repeated, only this time each sample was coded with a secret number so neither participants nor practitioners knew which was which. The link between the ability to resist and whether the volunteer was given the 'good' or 'bad' sugar disappeared. The results were no more than those that would be expected by chance. 'When these results were announced,' recalls Hyman, 'the head chiropractor turned to me and said, "You see, that is why we never do double blind testing any more. It never works" . . . since he "knew" that applied kinesiology works, and the best scientific method shows that it does not work, then – in his mind – there must be something wrong with the scientific method'.[55]

One way of thinking about these issues is to visualise CAM methods on a continuum of their scientific standing, determined by both their proven effectiveness and the scientific plausibility of their underlying principles. So at one end of the spectrum are the small number of CAM methods whose suggested mechanisms are both consistent with scientific knowledge and which have proven effects, such as a few herbal remedies and some relaxation techniques.

These are likely to be adopted by mainstream medicine if further research shows them to be both safe and effective. Somewhere in the middle are those like acupuncture: even though there is no scientific evidence for the existence of Qi energy, there are some demonstrable consequences of acupuncture treatment – relief of certain kinds of pain and nausea – which can be explained in scientific terms.[56] But the far end of the line is where we find the majority of CAM methods. Not only are therapies and diagnostic methods at this densely populated end of the continuum likely to be unproven or disproven by research, but their theoretical basis contradicts all current understanding of physics, anatomy, physiology, chemistry and biochemistry. Here we find homeopathy, reflexology, 'energy medicine' crystal and magnet therapy, distant healing, chiropractic and various alternative cancer treatments. Some approaches can appear more than once on this line, depending on what claims are being made; acupuncture, for example, reappears at this far end when it is promoted as a treatment for conditions other than pain or nausea.

Perhaps it is time to call a halt to research studies of those implausible practices at the extreme reaches of this line. Does continuing research bestow unwarranted attention, with their status being elevated simply by virtue of being investigated? So many of these studies end with the conclusion 'more research is needed', but how much more research of these unlikely therapies *do* we really need? American doctor Kimball C Atwood acknowledges that 'well-intentioned academic physicians may contend that the problem with many alternative medicine practices is that they have not been adequately studied'. But he thinks most are based on such implausible claims that to study them is a bad idea: 'It is bad science because the prior probability of their being correct is very close to zero; hence equivocal results – which are entirely expected in such cases – are paradoxically viewed as implying efficacy, or at least the need for more (equivocal) studies, on ad infinitum. It is bad public policy because it gives a scientifically naïve citizenry the misleading impression that legitimate scientists think such claims have merit, thus encouraging health fraud and waste'.[57]

The National Institute of Health and Clinical Excellence provides national guidance 'on promoting good health and preventing and treating ill health in the UK'. It does not evaluate complementary and alternative medicine, even though CAM is provided by the NHS. When the *British Medical Journal* published a debate in 2007 on whether NICE should evaluate alternative medicine it raised interesting questions. Those in favour, who included professor of nursing Linda Franck, said CAM deserves full evaluation by NICE because 'failure to evaluate complementary therapies leads to health inequalities because of uneven access and missed opportunities . . . if the evaluation is favourable, they should be adopted either on their own or integrated with conventional medicine'.[58] But professor of pharmacology and tireless quack buster David Colquhoun disagrees. He considers this would be a waste of time and money, because 'the evidence, such as it is, has been reviewed endlessly, and it is obvious that if NICE were to apply its normal criteria, almost all complementary and alternative medicine would be removed from the NHS immediately'. There was no point 'going through, yet again, evidence that we already know to be inadequate,' he wrote. NICE 'can't afford the time to do again what has already been done'.[59]

But in one last paradox, the integration of complementary medicine into the mainstream might just, in the end, be its undoing. Public health expert Angela Coulter has suggested persuasively that integration could undermine CAM's appeal, which 'lies in its dissimilarity from conventional medicine and the fact that complementary therapists are untrammelled by the bureaucracies of public health systems'. So if complementary methods became conventional, their distinctiveness, and hence their appeal, might well be reduced. She concludes that 'progress in this arena must be judged by the extent to which proponents of complementary medicine are willing to accept evaluation and regulation and the extent to which orthodox practitioners are prepared to improve their interpersonal skills'.[60]

We all, as consumers of health care, can reasonably expect medical science to develop ways to assess the effects of what it does. We require it to get its house in order and make necessary, evidence-based changes to its practice to give us the best chance of recovery when we are ill. We are entitled to have these same expectations of complementary and alternative medicine. Just like scientific medicine, CAM should be held accountable for what it says and does.

Alternative health practices should have no case for integration into the mainstream unless and until they have been examined under the microscope of scientific method and proven effective. By this standard homeopathy has no place in evidence-based health care and should not be provided by the NHS. Alternative medicine should also be subject to the same strictures as the pharmaceutical industry and not be allowed to sidestep regulation by calling its medicaments 'food supplements'. The Cancer Act must be more frequently enforced and revised to protect the sick and desperate from greedy cancer quacks. And our universities should not receive government subsidies to enable them to teach supernatural nonsense masquerading as science.

For any of this to happen those who use CAM will have to start thinking critically. We have seen how alternative medicine lives or dies by the market and in the end its future will be determined by its consumers. They must stop passively accepting endless absurd claims and understand that alternative medicine, and its practitioners, should be subject to rigorous interrogation, regulation and accountability. So, finally, another definition. *Sceptic:* one who demands reasonable evidence and logical justification before granting provisional assent to truth claims.[61] Don't be fooled again.

Acknowledgements

I would like to thank my husband Sam Organ for all the help he has given me in the writing of this book. I am grateful to our daughters Isabel and Judith Shapiro, who kept me going more than they realise. Special thanks are also due to my agent James Macdonald Lockhart who has provided constant guidance and wise advice. I am indebted to Stuart Williams at Harvill Secker who has been an inspiring and exacting editor.

Thanks for assistance and encouragement also go to Maria Duggan, Angharad Penrhyn Jones, Richard Stein, Professor Michael Baum, Robyn Sisman, Dr Harriet Hall, Shelley Charlesworth, Rose Jennings, Sarah Culshaw, Fiona Trier, Nicci Gerrard, Rosie Spurr, Stephen Loach, Alison Steele, Ruth Günay, Philippa Lowthorpe, Simon Jennings, Hilary Grieve, Stephanie Calman, the Healthfraud discussion list, George Monbiot, Adam Sisman, Jenny Moultrie, Professor David Colquhoun, Peter Moran, Rina Vergano, Ellie Steel, Michael Morris, Glenys Rowe and Pam Ryan-Ross.

Further Reading

Snake Oil Science, R Barker Bausell, Oxford University Press, 2007
The Skeptic's Dictionary, Robert Todd Carroll, Wiley, 2003
The Whole Truth, Ros Coward, Faber and Faber, 1989
Snake Oil and Other Preoccupations, John Diamond, Vintage, 2001
Bad Science, Ben Goldacre, 4th Estate, 2008
Bizarre Beliefs, Simon Hoggart and Mike Hutchinson, Prometheus Books, 1997
Natural Causes, Dan Hurley, Broadway Books, 2006
Voodoo Science, Robert Park, Oxford University Press, 2000
The Greatest Benefit to Mankind, Roy Porter, HarperCollins, 1997
Why People Believe Weird Things, Michael Shermer, Owl Books, 1997
Trick or Treatment, Simon Singh & Edzard Ernst, Bantam Press, 2008
Hippocratic Oaths, Raymond Tallis, Atlantic Books, 2004
Counterknowledge, Damian Thompson, Atlantic Books, 2008
Bad Medicine, Christopher Wanjek, Wiley, 2002
How Mumbo Jumbo Conquered the World, Francis Wheen, Fourth Estate, 2004

Websites

Bandolier: evidence-based thinking about medicine including reviews, analyses and information on alternative medicine.
jr2.ox.ac.uk/bandolier/booth/booths/altmed

Ebm-first: sceptical information on alternative medicine in general and links to research articles.
www.ebm-first.com

DC's Improbable Science: Professor David Colquhoun on truth, falsehood and evidence, including investigations of dubious and dishonest science.
www.dcscience.net

Healthwatch UK: assessment and testing of treatments as well as information on why only thoroughly tested treatments are safe.
www.healthwatch-uk.org

Quackwatch: non-profit corporation which combats health-related frauds, myths, fads and fallacies. Information on quackery, questionable therapies and products; linked to many other useful sites such as Homeowatch and Chirobase.
www.quackwatch.org

Sense about Science: independent organisation which promotes good science and evidence for the public.
www.senseaboutscience.org.uk

The Skeptics Dictionary: definitions and essays on alternative medicine, ideas and practices, regularly updated.
www.skepdic.com

The Quackometer: automated quack medicine measurement program – just enter the name of any website and the Quackometer will test it for quackery, returning immediate results.
www.quackometer.net

James Randi Educational Foundation: promotes critical thinking, includes online Encyclopaedia of Claims, Frauds & Hoaxes of the Occult and Supernatural.
www.randi.org

Science-Based Medicine:
Exploring issues and controversies in the relationship between science and medicine.
www.sciencebasedmedicine.org

Picture Credits

Notes

Preface and Chapter 1: Ancient and Modern

1 Christopher Smallwood et al., 'The Role of Complementary Medicine in the NHS', Fresh Minds for The Prince's Foundation for Integrated Health, 2005
2 'Complementary Medicines – UK', Mintel International Group, 2005
3 'Complementary Medicines – UK', Mintel International Group, 2007
4 K Thomas et al., 'Trends in access to complementary or alternative medicines via primary care in England: 1995–2001 Results from a follow-up national survey', *Family Practice*, 2003
5 'Chinese Herbal Medicine', Five Live Report BBC Radio 5 Live, 2005
6 *Psychic Surgeon* BBC 2 TV documentary, Angharad and Sara Penrhyn Jones, 2005
7 Professor Martin Knapp and Professor Martin Prince, 'Dementia UK, report into the prevalence and cost of dementia', London School of Economics and Alzheimer's Society, 2007
8 'Circadian Balance', Apollo Health, apollolight.com, 2007
9 lifepositive.com, 2007
10 David Adam, 'Scientists create GM mosquitoes to fight malaria and save thousands of lives', the *Guardian*, 2005
11 Roy Porter, *The Greatest Benefit to Mankind: a medical history of humanity from antiquity to the present*, HarperCollins, 1997
12 CNN news report, 1998
13 Jack Raso, *Dictionary of Metaphysical Healthcare*, Georgia Council Against Health Fraud, 1997
14 Michelle Roberts, 'What caused Gwyneth's spots', BBC News, 2004
15 Penelope Ody, *Secrets of Chinese Herbal Medicine*, Dorling Kindersley, 2001
16 David L Sackett et al., 'Evidence based medicine: What it is and what it isn't', *British Medical Journal*, 1996
17 James C Whorton, *Nature Cures – The history of alternative medicine in America*, OUP, 2002
18 Raymond Massey, 'Quack Cures', US Food and Drug Administration, 1959
19 James C Whorton, *Nature Cures – The history of alternative medicine in America*, OUP, 2002
20 Abraham Flexner, 'Medical Education in the United States and Canada', The Carnegie Foundation for the Advancement of Teaching, 1910
21 Adam Curtis, *The Century of the Self*, BBC TV documentary series, 2002
22 Arthur J Barsky, *Worried Sick: Our Troubled Quest for Wellness*, Little, Brown, 1988
23 Andrew Weil, *Health and Healing*, Houghton Mifflin Co., 1988
24 WHO strategy for traditional medicine, 2002
25 A Molassiotis et al., 'Use of complementary and alternative medicine in cancer patients: a European survey', *Annals of Oncology*, 2005
26 M A Richardson et al., 'Complementary/alternative medicine use in a comprehensive cancer center and the implications for oncology', *Journal of Clinical Oncology*, 2000

27 David M Eisenberg et al., 'Trends in alternative medicine use in the United States', *Journal of the American Medical Association*, 1998

28 Victor Ong, 'Classical Acupuncture in South Staffordshire', acunpunctureforlife.co.uk, 2007

29 Mark Lawson, 'Feeding the fear gene', the *Guardian*, 2007

30 E Ernst & A White BBC Survey of Complementary Medicine Use, 2000

31 Dr Ben Goldacre, 'Gillian McKeith, round 2', Bad Science, the *Guardian*, 2004

32 Fidelma Cook, 'Is Channel 4's latest food guru really a Quack and a danger to our health?' *Mail on Sunday*, 2004

33 'Backing for crackdown on bogus alternative medical practitioners', UK Department of Health press release, 2005

34 The Prince of Wales' speech at launch of The Prince's Foundation for Integrated Health's Five Year Strategy, 2003

35 The Prince of Wales' speech at launch of The Prince's Foundation for Integrated Health GP Associates, 2005

36 Christopher Smallwood et al., 'The Role of Complementary Medicine in the NHS', Fresh Minds for The Prince's Foundation for Integrated Health, 2005

37 Mark Henderson, 'Prince plots alternative treatments for the NHS', *The Times*, 2005

38 Richard Horton, letter to the *Guardian*, 2005

39 Catherine Zollman and Andrew Vickers, 'Complementary Medicine and the Patient', *British Medical Journal*, 1999

40 British Complementary Medicine Association, bcma.co.uk, 2005

41 'Homeopathy Awareness Week', Society of Homeopaths, 2007

42 Michael Baum, 'An open letter to the Prince of Wales: with respect, your Highness, you've got it wrong', *British Medical Journal*, 2004

43 'Improbable Science' page, dcscience.net/improbable.html, 2007

44 John Diamond, *Snake Oil and Other Preoccupations*, Vintage, 2001

45 badscience.net, 2007

46 Quackwatch Mission Statement, Quackwatch.com, 2007

47 Stephen Barrett, *Review of Complementary and Alternative Medicine in the United States*, Medscape General Medicine, 2005

48 James Randi's million dollar challenge, Randi.org, 2007

49 Dr Yasuhiro Suzuki, WHO Executive Director for Health Technology and Pharmaceuticals, WHO statement, 2002

Chapter 2: How to Spot a Quack

1 Definitions of 'quack' and 'kwakzalver', *Shorter Oxford English Dictionary on Historical Principles*, Clarendon Press, 1978

2 Stephen Barrett, 'Quackery: how should it be defined?' Quackwatch.com, 2001

3 Steven F Hotze MD, speaking on KSEV radio, Houston, US, 2006

4 Dr David Dowson, Avon Complementary Medicine UK, 2006

5 'Angel Guidance', Angel House Doctor service, angelguidance. co.uk, 2006

6 American Academy of Acupuncture, aaaom.edu, 2006

7 Maharishi Ayurveda, Maharishi.co.uk, 2006

8 Cass and Janie Jackson, *The Healing Power of Crystals*, Caxton Publishing, 2001

9 *Reiki,* Natural Health Service London Clinic, naturalhealthservice.org, 2006

10 Prof Rory Coker, 'Distinguishing Science and Pseudoscience', Quackwatch.org, 2001

11 Steven Novella, 'Quacks use quantum mechanics to make themselves look smarter', *New Haven Advocate*, 2006

12 'Experience Festival' website, 2005

13 Quantumtouch.com, 2006

14 'Allergy testing using the quantum medical system', Hale Clinic, haleclinic.com, 2007

15 Barefoot doctor website's terms and conditions, barefootdoctorglobal.com, 2005

16 Paul Lennard, Craniosacral Therapy, Welbeck-Pure Clinic, activeingredient.co.uk, 2006

17 D D Palmer, *The Science, Art and Philosophy of Chiropractic*, Portland, Oregon: Portland Printing House Company, 1910

18 *Subluxations*, patient information leaflet, Patient Media Inc., 2002

19 Hulda Regehr Clark, *The Cure for All Diseases*, B Jain publishers, 1999

20 Hulda Regehr Clark, *The Cure for All Cancers*, B Jain publishers, 1993

21 Jordan Rubin and David Remedios, *The Great Physician's Rx for Health and Wellness: seven keys to unlock your health potential*, Thomas Nelson Books, 2006

22 Penelope Ody, *The Secrets of Chinese Herbal Medicine*, Dorling Kindersley, 2001

23 Deepak Chopra, Ayurveda – the science of life, chopra.com, 2007

24 jontimayer.com, 2006

25 Jurak Classic Whole Body Tonic information, 2004

26 Shirley's wellness café, shirleys-wellness-café.com, 2006

27 Vithoulkas et al., 'Aggravations of the Symptoms After the Indicated Homeopathic Remedy', Hahnemannian Gleanings, Calcutta, 1978

28 Homeopathic Online Education – A cyberspace academy, simillimum.com, 2006

29 Jay W Shelton, *Homeopathy – how it really works*, Prometheus Books, 2004

30 S Grabia and E Ernst, 'Homeopathic aggravations: a systematic review of randomised, placebo-controlled clinical trials', *Homeopathy: the Journal of the Faculty of Homeopathy*, 2003

31 What Doctors Don't Tell You, wddty.com, 2006

32 Kevin Trudeau, naturalcures.com, 2006

33 cancertutor.com, 2006

34 chemotherapynews.com, 2006

35 Grounding Mattress Pad, Sleep and Wellness News, Rejuvinex™, 2006

36 Nature's Sunshine CleanStart™ 14-Day Cleanse advertisement 2006

37 Maureen Rooney, *Serrapetase News*, 2005

38 Madé Delaveris and Kate Sutton, 'Sutton's Solutions Colloidal Silver', booklet, 2001

39 Dr Lorraine Day Power Plan for Health – rebuild your immune system naturally without drugs, drday.com, 2006

40 Angharad and Sara Penrhyn Jones, *Psychic Surgeon*, BBC2 TV, 2005

41 Kathryn Flett, 'Off the rails', *Observer* newspaper, 2005

42 robynwelch.com, 2006

43 Angharad and Sara Penrhyn Jones, *Psychic Surgeon*, BBC2 TV, 2005

44 About Brandon Bays, The Journey, thejourney.com, 2007

45 Patient testimonial, drrind.com, 2006

46 Cancell effectiveness study, alternativecancer.us, 2006

47 Patient information, Raymond Perrin, Osteopath, Manchester UK, 2006

48 Most astonishing health disaster of the twentieth century, mercola.com, 2007

49 Belinda Grant Viagas, *Nature Cure – A Guide to Healthy Living*, Newleaf, 1999

50 Belinda Grant Viagas, *Stress – Restoring Balance to our Lives*, The Women's Press, 2001

51 'FTC targets bogus claims for pills advertised to make kids taller', press release, US Federal Trade Commission, 2006

52 Dr Joanna M Hoeller, Precision Spinal Care, Seattle, psbl.com/hoeller/, 2007

53 Carol Dale, 'Reflexology, Thermo-Auricular Therapy, Indian Head Massage & Reiki', Practice leaflet, 2006

54 Jeff Stratton, 'Far from benign – is South Florida's most notorious pimp peddling a fake cancer cure?' *Broward-Palm Beach New Times,* newtimesbpb.com, 2005

55 Alternative Cancer Treatment Manufacturers, alternativecancer.us, 2007

56 Professor Robert Park, 'The seven warning signs of bogus science', *The Chronicle of Higher Education,* 2003

57 lynnehancher.co.uk, 2006

58 What to expect during an ear candling session, by Cheryl Ashby, colon-health.net 2006

59 Testimonials, from ear-candles.net, 2006

60 D R Seely et al., 'Ear Candles: Efficacy and Safety', *Laryngoscope,* 1996.

61 Richard Harris, Ph.D.; Brigham Young University; Provo, UT posted on internet Audiology Forum, rcsullivan.com, 1999

62 'It's your health – ear candling', HealthCanada, hc-sc.gc.ca, 2005

63 Lisa M L Dryer, 'Why ear candling is not a good idea', Quackwatch.com, 2005

64 Philip Kaushall and Justin Nevill Kaushall, 'On Ear Cones and Candles', *Skeptical Enquirer,* 2000

65 Discover Pure Wellness, Biosun, Biosun.de, 2007

66 Elizabeth Bromstein, 'Wax on, wax off – does candling clear canal or burn it?' *Now Magazine,* nowtoronto.com, 2005

Chapter 3: Full of Eastern Promise

1 James Reston, 'Now, let me tell you about my appendectomy in Peking', *The New York Times,* 1971

2 John Adams on *Nixon in China,* John Adams official website, earbox.com, 2000

3 James C Whorton, *Nature Cures – the history of alternative medicine in America,* Oxford University Press, 2002

4 Dr Walter Tkach, reported in *Today's Health,* 1972

5 Patient information material, China Acupuncture Health Centre, Westford, Massachusetts, USA, 2006

6 Roy Porter, *The Greatest Benefit to Mankind,* HarperCollins, 1997

7 R H Bannerman, *Acupuncture: the WHO View,* World Health, 1979

8 Dr Paul Unschuld, *Medicine in China: a History of Ideas,* Berkeley: University of California Press, 1985

9 Penelope Ody, *The Secrets of Chinese Herbal Medicine,* Dorling Kindersley, 2001

10 Robert Imrie, 'Acupuncture: the facts', powerpoint presentation available at drspinello.com, 2006

11 J R Worsley, *Traditional Chinese Acupuncture, Meridians and Points,* Salisbury: Element Books, 1982

12 Felix Mann, 'A New System of Acupuncture', in Filshie and White's *Medical Acupuncture: a Western Scientific Approach,* Churchill Livingstone, 1998

13 Interview with Dr Paul Unschuld, *Acupuncture Today,* 2004

14 Interview with Dr Paul Unschuld, *Acupuncture Today,* 2004

15 D W Y Kwok, *Scientism in Chinese Thought,* Yale University Press, 1965

16 Quoted in Paul Unschuld, *Medicine in China: a History of Ideas,* University of California Press, 1985

17 Kim Taylor, *Chinese Medicine in Early Communist China,* Routledge Curzon, 2005

18 Zhisui Li, *The Private Life of Chairman Mao: The inside story of the man who made modern China*, Chatto and Windus, 1994

19 Zhu Lian, 2nd edition of *The New Acupuncture*, People's Medical Publishers, 1954

20 Kurt Butler, *A Consumer's Guide to 'Alternative' Medicine*, Prometheus Books, 1992

21 *Barefoot Doctor's Manual*, English translation of the official Chinese paramedical textbook, Running Press, 1977

22 Mark Magnier, 'Chin's Medicine Wars', *Los Angeles Times*, 2007

23 'Traditional Chinese medicine losing out to western drugs: online survey', People's Daily online, english.people.com, 2006

24 Nicholas Zamiska, 'On the Trail of Ancient Cures – Swiss Drug Maker Novartis Looks to Chinese Medicine in its Search for New Products', *Wall Street Journal*, 2006

25 Marianne Barriaux, 'Merck looks to traditional remedies with Chi-Med deal', the *Guardian*, 2006

26 K Streitberger and J Kleinheinz, 'Introducing a placebo needle into acupuncture research', *Lancet*, 1998

27 Jongbae Park et al., *Acupuncture in Medicine*, Vol 20, 2002

28 Mark Aird et al., 'A Study of the Relative Precision of Acupoint Location Methods', *Journal of Alternative and Complementary Medicine*, 2002

29 Andrew Vickers et al., 'Do certain countries produce only positive results – a systematic review of controlled trials', *Controlled Clinical Trials* journal, 1998

30 Kurt Butler, *A Consumer's Guide to Alternative Medicine*, Prometheus Books, 1992

31 H C Keng and N H Tao, 1985. Translated by Paul Unschuld. 'The evaluation of acupuncture anaesthesia must seek truth from facts'. In *Medicine in China: a History of Ideas*, by P U Unschuld, Berkeley: University of California Press

32 Stephen Barrett MD and William Jarvis PhD, *The Health Robbers: A Close Look at Quackery in America*, Prometheus Books, 1993

33 Barry Beyerstein and Wallace Sampson, 'Traditional Medicine and Pseudoscience in China', *Skeptical Inquirer*, 1996

34 National Council Against Health Fraud position paper on Acupuncture, 1999

35 Dr Paul Unschuld, *Chinese Medicine,* Paradigm Publications, 1998

36 Ian Bell, 'As they're wont to say in Edinburgh, you'll have had your chi', *The Herald*, 2006

37 Simon Singh, 'Did we really witness the "amazing power" of acupuncture?' *The Daily Telegraph*, 2006

38 *Alternative Medicine: The Evidence*, BBC/Open University 2006

39 Louisa Lim, *The high price of illness in China*, BBC News 2006

40 Francis Markus, *China's ailing healthcare*, BBC News, 2004

41 'Use of Traditional Medicine in the Western Pacific Region', WHO, 2005

42 Thomas J Wheeler PhD, *A Scientific Look at Alternative Medicine*, University of Louisville School of Medicine, 2005

43 jr2.ox.ac.uk/bandolier, 2007

44 'Effective Health Care: Acupuncture', NHS Centre for Reviews and Dissemination, University of York, Royal Society of Medicine Press, 2001

45 Stephen Barrett MD, 'Acupuncture, Qigong and Chinese Medicine', Quackwatch.org, 2004

46 James Reston, ibid.

47 Reyhan Harmanci, 'Healthy Doubts', sfgate.com, 2006

48 Professor G Ulett, J Han and S Han, 'Traditional and evidence-based acupuncture: History, mechanisms, and present status', *Southern Medical Journal*, 1998

49 Professor G Ulett, 'Acupuncture: another view', letter to *Psychiatric News*, Journal of the American Psychiatric Association, 2002

50 E Ernst and A White, 'Life Threatening Adverse Reactions after Acupuncture? A systematic review', *Pain*, 1997

51 E Ernst et al., 'Adverse events following acupuncture: prospective survey of 32000 consultations', *BMJ*, 2001

52 Bryan L Frank and Nader Soliman, 'Shen Men: a critical assessment through advanced auricular therapy', *Medical Acupuncture Physicians' Journal*, 1998

53 George A Ulett and SongPin Han, *The Biology of Acupuncture*, Warren H Green Inc, 2002

54 Biography of Eunice Ingham, reflexology-uk.co.uk

55 A R White et al., 'A blinded investigation into the accuracy of reflexology', *Complementary Therapy Medicine*, 2000

56 Carol Dale, 'Reflexology, Thermo-Auricular Therapy, Indian Head Massage & Reiki', practice leaflet, 2006

57 'What is the history of Reiki?' reiki.org, 2006

58 'Complementary Medicine, A Guide for Patients', Prince's Foundation for Integrated Health, 2005

59 reikiwithtrust.com, 2006

60 Penelope Ody, *The Secrets of Chinese Herbal Medicine*, Dorling Kindersley, 2001

61 Register of Chinese Herbal Medicine, rchm.co.uk, 2006

62 Penelope Ody, *The Secrets of Chinese Herbal Medicine*, Dorling Kindersley, 2001

63 Roy Alder, Dept. of Health International Industry Division, speaking at 2nd International Science and Technology Conference on TCM, 2005

64 Register of Chinese Herbal Medicine, rchm.co.uk, 2007

65 E Ernst and N C Armstrong, 'The treatment of eczema with Chinese herbs: a systematic review of randomised clinical trials', *British Journal of Clinical Pharmacology*, 1999

66 Barney Calman, 'The Chinese Medicine Minefield', *Daily Mail*, 2007

67 Herbal Information Centre, kcweb.com, 2006

68 Marilyn Glenville, *Natural Health* website, marilynglenville.com

69 Entry for Dong Quai/Angelicae gigantis radix, Natural Medicines Comprehensive Database 2006, Therapeutic Research Facility, 2006

70 E Ernst, 'Contamination of herbal medicines', *Pharmaceutical Journal*, 2005

71 'FDA Plans Regulation Prohibiting Sale of Ephedra-Containing Dietary Supplements and Advises Consumers to Stop Using These Products', US Food and Drugs Administration Consumer Alert, 2003

72 Herbalean, herba-lean.com, 2007

73 *Gold star nutrition – Power Thin w/ephedra original blend*, luckyvitamin.com, 200z

74 Richard J Ko, 'Adulterants in Asian Patent Medicines', *New England Journal of Medicine*, 1998

75 'Health Canada warns consumers not to take the Chinese medicine Shortclean due to potential health risk', Health Canada press release, 2005

76 'Making treatment decisions: Chinese Herbal Medicine', American Cancer Society, 2006

77 Edzard Ernst, 'Adulteration of Chinese Herbal Medicines with synthetic drugs: a systematic review', *Journal of Internal Medicine*, 2002

78 Dr Helen Ramsay et al., 'Herbal creams used for atopic eczema in Birmingham,

UK illegally contain potent corticosteroids', Archives of Disease In Childhood, 2003

79 *Chinese medicine outlets probed*, BBC Radio 5 Live report, 2005

80 'Public Statement on the Risks associated with the use of herbal products containing the Aristolochia species', European Medicines Agency, 2005

81 Prof John Louis Vanherweghem et al., 'Rapidly Progressive Interstitial Renal Fibrosis in Young Women: association with slimming regimen using Chinese herbs', *The Lancet*, 1997

82 Severin Carrell, 'Alarm over huge increase in "dangerous" herbal remedies', *Independent on Sunday*, 2006

83 'Traditional Chinese Medicines', Medicines and Healthcare products Regulatory Agency, 2006

84 Department of Ayurveda, Unani, Siddha and Homeopathy; Ministry of Health & Family Welfare, Govt of India, 2006

85 'What is Ayurveda?' ayurbalance.com

86 'What is Ayurvedic Medicine?' National Centre for Complementary and Alternative Medicine, National Institutes of Health, US Department of Health 2005

87 R B Saper et al., 'Heavy metal content of Ayurvedic herbal medicine products', *JAMA*, 2004

88 Edzard Ernst, 'Heavy metals in traditional Indian remedies', *European Journal of Clinical Pharmacology*, 2002

89 Alliance for a New Humanity, anhglobal.org

90 Deepak Chopra, *Perfect Health*, Bantam, 2001

91 chopra.com

92 torsparetreat.com

93 Gita Mehta, *Karma Cola: Marketing the Mystic East*, Jonathan Cape, 1979

Chapter 4: Eurotrash

1 Egger et al., 'Are the clinical effects of homoeopathy placebo effects? Comparative study of placebo-controlled trials of homoeopathy and allopathy', *The Lancet*, 2005

2 'The End of Homoeopathy', *The Lancet*, 2005

3 R Medhurst, 'The Use of Homeopathy around the World', *Journal of the Australian Traditional-Medicine Society*, 2004

4 H Ni et al., Utilisation of Complementary and Alternative Medicine by United States Adults: Results from the 1999 National Health Interview Survey, *Medical Care*, 2002

5 PEK study, Swiss Federal Office of Public Health, 2005

6 E Ernst, 'Is homeopathy a clinically valuable approach?' *Trends in Pharmacological Science*, 2005

7 Denis Campbell and Mary Fitzgerald, 'Royals' favoured hospital at risk as homeopathy backlash gathers pace', *Observer*, 2007

8 Royal Insight, royal.gov.uk, 2006

9 'International network for the history of hospitals newsletter', European Association for the History of Medicine paper presented at the International Foundation for Homeopathy, Seattle USA Medicine and Health, 2002

10 Francis Treuherz, 'Homeopathy Around the World', *The Homeopath*, 1992

11 E Ernst, 'Is homeopathy a clinically valuable approach?' *Trends in Pharmacological Science*, 2005

12 Francis Treuherz, 'Homeopathy Around the World', *The Homeopath*, 1992

13 David S Spence et al., 'Homeopathic Treatment for Chronic Disease: A 6-year

University Hospital Outpatient Observational Study', *Journal of Alternative and Complementary Medicine*, 2005

14 Samuel Hahnemann, 1842 edition of the *Organon of Medicine*, edited and annotated by Wenda Brewster O'Reilly, Birdcage Books, 1996

15 Manesh Bhatia, homeopathy 4 everyone, hpathy.com, 2006

16 Dr William F Thomas MD, Hahnemann's allergy to quinine, angelfire.com, 2006

17 Louis Klein, Hahnemannian Proving of Argentum Sulphuricum, homeopathycourses.com, 2006

18 George Vithoulkas, *The Science of Homeopathy*, Grove Press, 1980

19 Jay W Shelton, *Homeopathy: how it really works*, Prometheus Books, 2004

20 'About Proving Methods', Dynamis School of Advanced Homeopathic Studies, 2006

21 Mary English, Tempesta: The homeopathic proving of Tempesta the Storm, veryscarymary/stormremedy, 2006

22 Charles Wansborough, The Berlin Wall: A Remedy of Power, biolumanetics.net, 2006

23 Samuel Hahnemann, 1842 edition of the *Organon of Medicine*, edited and annotated by Wenda Brewster O'Reilly, Birdcage Books, 1996

24 Homeopathy Manufacturing, Institut Boiron, boiron.com, 2006

25 Robert Park, *Voodoo Science*, Oxford University Press, 2000

26 Beth MacEoin, *Homeopathy – the practical guide for the 21st century*, Kyle Cathie Limited, 2006

27 MRC Hans, Homeopathic dilution rates, otherhealth.com, 2005

28 Jan Willem Nienhuys, 'The True Story of Oscillococcinum', *Skepter Magazine*, 1994

29 Robert Park, *Voodoo Science*, Oxford University Press, 2000

30 Dan McGraw, 'Flu symptoms? Try Duck', US News and World Report, 1997

31 Stephen Fulder, *The Handbook of Complementary Medicine*, Oxford University Press, 1990

32 Ronald Steriti, NMD, PhD, Homeopathy for highly sensitive people, naturdoctor.com, 2006

33 Holistic Health – homeopathy, shirleys-wellness-café.com, 2006

34 William Boericke, *New Manual of Homoeopathic Materia Medica and Repertory*, B Jain, 2000

35 Homeopathic library, internethealthlibrary.com, 2006

36 Beth MacEoin, *Homeopathy – the practical guide for the 21st Century*, Kyle Cathie, 2006

37 www.vitacost.com

38 Stephen Fulder, *Handbook of Complementary Medicine*, Oxford University Press, 1990

39 Ibid.

40 Thomas J Wheeler, 'A scientific look at Alternative Medicine – course notes', Department of Biochemistry and Molecular Biology, University School of Medicine, 2005

41 Beth MacEoin, *Homeopathy – the practical guide for the 21st Century*, Kyle Cathie, 2006

42 Beth MacEoin, *Homeopathy – the practical guide for the 21st Century*, Kyle Cathie, 2006

43 Laboratoires Boiron, Consolidated Results, Report to Shareholders, 2006

44 Products, Nelsons Homeopathy, nelsonshomeopathy.co.uk, 2007

45 Media Release, British Pharmacological Society, 2006

46 Medicines for Human Use (National Rules for Homeopathic Products) Regulations 2006, Lords Hansard, publications. parliament.uk, 2006

47 K Schmidt and E Ernst, 'MMR vaccination advice over the internet', *Vaccine*, 2003

48 Malaria and Homeopathy, Sense about Science, senseaboutscience.org.uk, 2006

49 Peter Fisher, *Homeopathy*, National Eczema Society Factsheet, National Eczema Society, 2001

50 Robin Shohet, *Passionate Medicine*, Jessica Kingsley Publishers, 2005

51 Author's interview with Dr Trevor Thompson, Bristol Homeopathic Hospital, 2006

52 Dr Ben Goldacre and Dr David Spence radio discussion, *You and Yours* BBC Radio 4, 2006

53 Bulletin on the effectiveness of health service interventions for decision makers: Homeopathy, NHS Centre for Reviews and Dissemination, 2002

54 E Ernst, 'A systematic review of systematic reviews of homeopathy', *British Journal of Pharmacology*, 2002

55 Klaus Linde and Dieter Melchart, *Journal of Alternative and Complementary Medicine*, 1998, quoted in Thomas J Wheeler, 'Scientific look at Alternative Medicine' course notes, University of Louisville School of Medicine, 2005

56 P Fisher and D L Scott, 'A randomised controlled trial of homeopathy in rheumatoid arthritis', *Rheumatology*, 2001

57 *Homeopathy – the test*, Horizon, BBC2 TV, 2002

58 Jay Shelton, letter to the author, 2006

59 Caroline Richmond, 'Jacques Benveniste: obituary', the *Guardian* newspaper, 2004

60 Simon Hoggart, 'Healthy dollop of grease and a damp squib', the *Guardian* newspaper, 2006

61 Jane Seymour, 'As if by Magic', *New Scientist*, 2001

62 Jay Shelton, letter to the author, 2006

63 William T Jarvis PhD, Misuse of the Term 'Allopathy', National Council of Health Fraud, 2000

64 James C Whorton, *Nature Cures – the History of Alternative Medicine in America*, Oxford University Press, 2002

65 David W Cathell and William T Cathell, on *The Physician Himself*, Davis Company, Baltimore, Maryland, 1902

66 bachcentre.com, 2007

67 Alcohol-free recovery plus emergency spray, product details, Ainsworths, ainsworths.com, 2007

68 Philip M Chancellor, *Handbook of the Bach Flower Remedies*, C W Daniel, 1980

69 Study the work of Dr Edward Bach at the original source, course information material, Bach Visitor and Education Centre, 2007

70 bachcentre.com

71 John Diamond, *Snake Oil and Other Preoccupations*, Vintage, 2001

72 Varro E Tyler, False Tenets of Paraherbalism, Quackwatch.org, 1999

73 Catherine Zollman and Andrew Vickers, *ABC of Complementary Medicine*, BMJ Books, 2000

74 National herb awareness week, BBC News 2006

75 Traditional Herbal Medicines Registration Scheme: frequently asked questions, MHRA 2006

76 Twelve supplements you should avoid, ConsumerReports.org

77 Varro E Tyler PhD, False Tenets of Paraherbalism, Quackwatch.org, 1999

78 E Ernst, 'Complementary Medicine – the evidence so far', Peninsula Medical School, University of Exeter, 2005

79 Charles Morris MD and Jerry Avorn MD, 'Internet Marketing of Herbal Products', *Journal of the American Medical Association*, 2003

80 M Walji et al., 'Efficacy of quality criteria to identify potentially harmful information: a cross-sectional survey of complementary and alternative medicine web sites', *Journal of Medical Internet Research*, 2004

81 Stephen Barrett, The Herbal Minefield, Quackwatch.org, 2004

82 Z-J Wang et al., 'Contents of hypericin and pseudohypericin in five commercial products of St John's wort (Hypericum perforatum)', *Journal of the Science of Food and Agriculture*, 2004

83 consumerlab.com, 2007

84 About Herbs database, Memorial Sloan-Kettering Cancer Center, mskcc.org/mskcc/html/11570.cfm

85 'St John's Wort And Echinacea Interfere With Some Drugs By Moving Them Out Of The Body Too Fast', Medicalnewstoday.com, 2006

86 A A Izzo and E Ernst, 'Interactions between herbal medicines and prescribed drugs: a systematic review', *Drugs*, 2001

87 M K Ang-Lee et al., 'Herbal medicines and perioperative care', *Journal of the American Medical Association*, 2001

88 S Bent et al., 'Saw Palmetto for Benign Prostatic Hyperplasia', *New England Journal of Medicine*, 2006

89 M C van Dongen et al., 'The efficacy of ginkgo for elderly people with dementia and age-associated memory impairment', *Journal of the American Geriatric Association*, 2000

90 D O Clegg et al., 'Glucosamine, chondroitin sulfate, and the two in combination for painful knee osteoarthritis', *New England Journal of Medicine*, 2006

91 Stephan Reichenbach et al., 'Meta-analysis: Chondroitin for osteoarthritis of the knee or hip', *Annals of Internal Medicine*, 2007

92 Carl Bartecchi, 'Herbs and Supplements – help or hype?' *The Pueblo Chieftain,* 2006

93 Katherine M Newton et al., 'Treatment of Vasomotor Symptoms of Menopause with Black Cohosh, Multibotanicals, Soy, Hormone Therapy, or Placebo', *Annals of Internal Medicine*, 2006

94 Black Cohosh, Natural Medicines Comprehensive database, 2006

Chapter 5: Junk Science

1 Carolyn Thomas de la Peña, *The Body Electric: How Strange Machines Built the Modern American*, New York University Press, 2003

2 Bob McCoy, *Quack! Tales of Medical Fraud from the Museum of Questionable Medical Devices*, Santa Monica Press, 2000

3 A vibration of energy taking us beyond the 21st century, Aquarian Healing International, 2006

4 gentlewindproject.org, 2006

5 'NH residents eligible for refunds from defunct Maine project', *Boston Globe*, 2007

6 Leonard Finegold and Bruce L Flamm, 'Magnet Therapy – Extraordinary claims, no proven benefits', *British Medical Journal*, 2006

7 Magnet Therapy, About Herbs, Botanicals and other products, Memorial Sloan Kettering Cancer Center, 2006

8 Robert Todd Carroll, 'Mesmerism', *The Skeptic's Dictionary*, Wiley, 2003

9 James D Livingston, 'Magnetic Therapy: Plausible Attraction', *Skeptical Inquirer*, 1998

10 Sales information for Green Foam insoles, bodykind.com 2006

11 Christopher Wanjek, *Bad Medicine*, John Wiley, 2003

12 Donald G McNeil Jr., 'Strong magnets cited in accidents', *New York Times*, 2005

13 Robert Park, *Voodoo Science*, Oxford University Press, 2000

14 Leonard Finegold and Bruce L Flamm, 'Magnet Therapy – Extraordinary Claims, no proven benefits', *British Medical Journal*, 2006

15 Robert Park, *Voodoo Science*, Oxford University Press, 2000

16 Sarah-Kate Templeton, 'NHS takes up Cherie's magic magnets cure', *The Sunday Times*, 2006

17 Dr Nyjon Eccles, 'Effects of 4UlcerCare on leg ulcer recurrence and the potential cost savings to the NHS', *Journal of Wound Care*, 2005

18 Professor David Colquhoun, Improbable Science, ucl.ac.uk/ Pharmacology/dc-bits/quack, 2006

19 OFT takes action and Magnopulse Limited agrees to give undertakings, Press Release, Office of Fair Trading, 2007

20 Epic – a technological breakthrough, Aixmed Gmbh, web.aixmed.de, 2007

21 EAV Discussions 'The basics', VeraDyne Corporation, veradyne.com, 2007

22 EAV overview and history, Biotechhealth.com, 2006

23 Meridian Stress Assessment System, electronichealing.co.uk, 2006

24 *Food Sensitivity, Inside Out*, BBC TV, 2003

25 Robert Mosenkis Examination of a Vegatest Device, Quackwatch, 2005

26 Stephen Barrett MD, Quack 'Electrodiagnostic' Devices, Quackwatch, 2005

27 Stephen Barrett MD, Quack 'Electrodiagnostic' Devices, Quackwatch, 2005

28 Azizah Clayton and John Kelsey, 'The Complete Package for the Quantum QXCI & SCIO', QQS, 2006

29 Michael J Behrens and Christine Willsmen, 'How one man's invention is part of a growing worldwide scam that snares the desperately ill', *Seattle Times*, 2007

30 'Quantum Scio', *Energy Medicine Review*, 2004

31 The QXCI device, energetic-medicine.net, 2006

32 'Quantum Scio', *Energy Medicine Review*, 2004

33 The complete package for the Quantum QXCI & SCIO, qxcisynergy.co.uk, 2007

34 Take your practice to the 'next level' of natural health care, NutriVital Health, Nutrivital.co.uk, 2007

35 Stephen Barrett MD, Dubious Diagnostic Tests, Quackwatch.org, 2006

36 Bionetics: its science, philosophy and history, Bionetics.co.uk, 2006

37 Entry for Pau D'Arco, Natural Medicines Comprehensive Database, 2005

38 Bionetics, ASA adjudication, asa.org.com, 2006

39 Mary Staggs Interview, detox-online.com, 2006

40 Hydra Detox at the Nightingale Clinic, The Nightingale Clinic, 2007

41 Detox effortlessly through your feet, press release, Hydra Detox International, 2004

42 Alex Howard, Why ME? alexhoward.me.uk, 2006

43 Water Color, IonSpa Manual, Wellspring Products, 2004

44 Chase Sammut Brain Tumour Blog, chasesammut.blogspot.com

45 What to expect from a session, IonSpa Manual, Wellspring Products, 2004

46 Bogus Cancer Cure Exposed, *Inside Out*, BBC TV, 2004

47 Carl Butler, 'Dying man was sold false cure', *Daily Post*, 2003

48 'Conman cancer therapist is jailed', BBC News, 2003

49 IFAS high frequency machines, ifas.com.au, 2006

50 David Hannaford, A short explanation of IFAS, Alumni Association of Natural
    Medicine Practitioners Inc, alumni.org.au, 2006

Chapter 6: Bad Backs

1   Sarah Bosely, 'We can rebuild you', the *Guardian*, 2006

2   Catherine Zollman and Andrew Vickers, *ABC of Complementary Medicine*, BMJ
    books, 2000

3   Catherine Zollman and Andrew Vickers, *ABC of Complementary Medicine*, BMJ
    books, 2000

4   Samuel Homola, *Chiropractic, Bonesetting, and Cultism,* online book, chirobase.org,
    1963 revised 2000

5   Thomas J Wheeler, 'A Scientific Look at Alternative Medicine course notes',
    University of Louisville School of Medicine, 2005

6   Prof E Ernst and P H Canter, 'A systematic review of systematic reviews of
    spinal manipulation', *Journal of the Royal Society of Medicine*, 2006

7   Dr Scott Kinkade, 'Evaluation and Treatment of Acute Low Back Pain', *Amer-
    ican Family Physician*, 2007

8   Stephen Paulus, Andrew Taylor Still (1828-1917), A Life Chronology of the
    First Osteopath, InterLinea.org, 2006

9   Andrew Taylor Still, *Autobiography of Andrew T Still with a History of the Discovery
    and Development of the Science of Osteopathy*, published by the author, 1897

10 From the curriculum of the Andrew Taylor Still University, Kirksville College
    of Osteopathic Medicine, atsu.edu, 2006

11 Stephen Paulus, Andrew Taylor Still (1828–1917), A life Chronology of the
    First Osteopath, InterLinea.org, 2006

12 W J J Assendelft et al., 'Spinal manipulative therapy for low-back pain', *The
    Cochrane Collaboration Reviews*, 2006

13 'Dr Palmer', *Davenport Leader*, 1894, quoted in Vern Gielow, *Old Dad Chiro: a
    biography of D.D. Palmer,* Bawden Brothers, 1981.

14 James C Whorton, *Nature Cures – the History of Alternative Medicine in America*,
    Oxford University Press, 2002

15 Daniel David Palmer, *The Science, Art and Philosophy of Chiropractic*, Portland
    Printing House Company, 1910

16 Daniel David Palmer, *A brief history of the Author and Chiropractic, The Chiro-
    practor's Adjuster*, Portland Printing House Company, 1910

17 Bobby Westbrooks, 'The troubled legacy of Harvey Lillard: the black experi-
    ence of chiropractic', *Chiropractic History*, 1982

18 Daniel David Palmer, *A brief history of the Author and Chiropractic, The Chiro-
    practor's Adjuster*, Portland Printing House Company, 1910

19 Daniel David Palmer, *A brief history of the Author and Chiropractic, The Chiro-
    practor's Adjuster*, Portland Printing House Company, 1910

20 Letter from D D Palmer to P W Johnson, 1910, Archives of David D Palmer,
    Health Sciences Library, Davenport, Iowa, USA.

21 B J Palmer, 'A Bit of History', *Fountainhead News*, 1919

22 Dr R V Pierce, *The people's common sense medical adviser in plain English, or,
    Medicine simplified*, Buffalo, New York, 1895

23 History of Palmer Chiropractic, Palmer College of Chiropractic,
    palmer.edu/pfch/PCCHistory.htm, 2006

24 Ralph Lee Smith, *At your own risk: the case against chiropractic*, Simon & Schuster,
    1969

25 Ralph Lee Smith, *At your own risk: the case against chiropractic*, Simon & Schuster, 1969

26 Ralph Lee Smith, *At your own risk: the case against chiropractic*, Simon & Schuster, 1969

27 Chiropractic Supply – diagnostic equipment, Nervoscope available from chirocity.com, 2006

28 Chiropractic History, Eastern Chiropractic, easternchiropractic. com, 2007

29 Joe Schwartz, 'Chiropractic', *Montreal Gazette*, 2000

30 The waiting list practice, teamwlp.com

31 C J Mertz, 'Make your procedures your promotions', *The Chiropractic Journal*, World Chiropractic Alliance, 1997

32 Anglo European College of Chiropractic, aecc.ac.uk, 2007

33 What is a subluxation? Echiropractic.net, 2006

34 M L Russell et al., 'Beliefs and Behaviours: understanding chiropractors and immunisation', *Vaccine*, 2004

35 Chiropractic Paradigm, Association of Chiropractic Colleges, chirocolleges.org, 2006

36 Consultation on Identity – Quantitative Research Findings, World Federation of Chiropractic, 2004

37 Who needs chiropractic? World Chiropractic Alliance, worldchiropracticalliance.org, 2006

38 Stephen Barrett MD, Chiropractic's Elusive Subluxation, Quackwatch, 2006

39 R Frogely, letter to Dr Stephen Barrett, quoted in Chiropractic's Elusive Subluxation, chirobase.org, 2006

40 Joseph C Keating Jr et al., 'Subluxation – Dogma or Science', *Chiropractic & Osteopathy*, 2005

41 Lesley Briggs et al., 'Measuring Philosophy: a philosophy index', *Journal of the Canadian Chiropractic Association*, 2002

42 'A different way to heal', *Scientific American Frontiers*, PBS, 2002

43 David Seaman, 'Philosophy and Science versus dogmatism in the practice of Chiropractic', *Journal of Chiropractic Humanities*, 1998

44 Samuel Homola DC, *Inside Chiropractic: A Patient's Guide*, Prometheus Books, 1999

45 Edzard Ernst, 'Chiropractor's use of X-rays', *British Journal of Radiology*, 1998

46 Consulting the Profession: A Survey of UK Chiropractic, General Chiropractic Council, 2004

47 Amy Berrington de Gonzalez and Sarah Darby, 'Risk of cancer from diagnostic X-rays: estimates for the UK and 14 other countries', *The Lancet*, 2004

48 X-rays – how safe are they? National Radiological Protection Board, 2001

49 Kurt Butler, *A Consumer's guide to alternative medicine*, Prometheus Books, 1992

50 Frances Denoon, interview with the author, 2006

51 Inquest into the death of Lana Dale Lewis, Coroner's jury verdict and recommendations, Office of the Chief Coroner, Ontario, gov.on.co/

52 Julian Branch, 'Chiropractor assured mother of dead woman she'd be ok', *The Canadian Press*, 1998

53 *Complementary Complications* (reporter Donogh Diamond), *Prime Time TV* programme, RTE, 2005

54 Dr Brad Stewart et al., Statement of Concern to the Canadian Public from Canadian Neurologists Regarding the Debilitating and Fatal Damage Manipulation of the Neck May Cause to the Nervous System, 2002

55 Michael Devitt, 'Ontario removes Chiropractic from Provincial Health Plan', *Dynamic Chiropractic*, 2004

56 Deanna M Rothwell et al., 'Chiropractic Manipulation and Stroke – a Population-based case-control study', *Stroke,* 2001

57 V Beletsky, 'Chiropractic manipulation may be underestimated as a cause of stroke', paper presented at a conference of the American Stroke Association, 2002

58 U Reuter et al., 'Vertebral artery dissections after chiropractic neck manipulation in Germany over three years', *Journal of Neurology,* 2006

59 Sunita Vohra, MD et al., 'Adverse Events Associated with Paediatric Spinal Manipulation: a systematic review', *Paediatrics,* 2006

60 Edzard Ernst, 'Spinal manipulation: Its safety is uncertain', *Canadian Medical Journal,* 2002

61 Chiropractic and Strokes: a balanced view, Press statement, General Chiropractic Council, 2002

62 Margaret A Munsey, What is NUCCA, psbl.com, 2006

63 Johanna Hoeller, Precision Spinal Care, psbl.com/hoeller

64 Harriet Hall MD, letter to Washington State Department of Health, 2006

65 Osteopathy in the Cranial field, osteodoc.com, 2006

66 Cranial sacral therapy/ SomatoEmotional release – Education for better patient care, International Alliance of Healthcare Educators, iahe.com, 2006

67 Press release for John E Upledger's autobiography 'Lessons Out of School: From Motown Gangs to Bar Jazz to Cell Talk – Life Stories', North Atlantic Books, 2006

68 What are the benefits of craniosacral therapy? The Craniosacral Association of the UK, 2006

69 Steve E Hartman and James M Norton, 'Interexaminer Reliability and Cranial Osteopathy', *The Scientific Review of Alternative Medicine,* 2002

70 P Sommerfeld et al., 'Inter and intraexaminer reliability in palpation of the "primary respiratory mechanism" within the "cranial concept"'. *Manual Therapy*

71 Mark E Rosen, Osteopathy in the Cranial Field, Osteopathy – Art of Practice, osteodoc.com, 2006

72 Brid Hehir, 'Head cases: an examination of craniosacral therapy', *Journal of the Royal College of Midwives,* 2003

73 Centre for cranio-sacral therapy, Primrose Hill Natural Health Centre, 2006

74 Bill Ferguson, Babies, billferguson.co.uk, 2006

75 Emma Mahony, 'It works for me: craniosacral therapy', *The Times,* 2005

76 Prospectus for two year professional training, Institute of Craniosacral Studies, 2006

77 Practitioner Trainings in biodynamic craniosacral therapy, The Craniosacral Therapy Educational Trust, 2006

78 Robert Todd Carroll, *The Skeptics Dictionary,* John Wiley, 2003

79 Leonard C Bruno, 'Therapeutic Touch', *Encyclopaedia of Medicine,* Gayl Research, 1999

80 The Therapeutic Touch Process, therapeutic-touch.org, 2006

81 Emily Rosa et al., 'A close look at therapeutic touch', *Journal of the American Medical Association,* 1998

82 William T London, letter to the editor, *Nurse Practitioner,* 2002

83 Rebecca Witner, *Hands that Heal: The Art of Therapeutic Touch,* Healing Arts, 1995

Chapter 7: Cancer

1 Katherine A See et al., 'Accidental death from acute selenium poisoning', *Medical Journal of Australia,* 2006

2 Prevalence of CAM use in the United Kingdom, Complementary and Alternative Medicine Assessment in the Cancer field, CAM-Cancer.org, 2006

3   Catherine Zollman and Andrew Vickers, 'What is Complementary Medicine?' *British Medical Journal*, 1999

4   E Y Chang et al., 'Outcomes of breast cancer in patients who use alternative therapies as primary treatment', *American Journal of Cancer*, 2006

5   CM McNeil et al., Delay in breast cancer diagnosis associated with ineffective alternative therapies. University of Sydney at Westmead Hospital, Westmead, NSW, Australia, 2006

6   T Risberg et al., 'Does use of alternative medicine predict survival from cancer?', *European Journal of Cancer*, 2003

7   Peter Moran, Alternative Medicine and Cancer, cancerwatcher. com

8   John Lennon, 'Whatever Gets You Through the Night', *Walls and Bridges* album, 1974

9   How to Build a Holistic Cancer Program, Cancer Options, canceroptions.co.uk, 2006

10   Denise Grady, 'Breast Cancer Study on Alternative Therapy', *New York Times*, 1999

11   Harol J Burstein et al., 'Use of Alternative Medicine by women with early-stage breast cancer', *New England Journal of Medicine*, 1999

12   Caroline Hoffman, interview with the author, 2006

13   How we can help you, Breast Cancer Haven DVD, 2006

14   Therapies, groups & classes, booklet, Breast Cancer Haven, 2006

15   How NES professional works – NES integrates physics and biology, nutrienergetics.com, 2006

16   Healing, pennybrohncancercare.org, 2006

17   Dr Rosy Daniel, *The Cancer Directory*, HarperThorsons 2005

18   Health Creation Training, healthcreation.co.uk, 2006

19   Tom Dyckhoff, 'Patients are a virtue', *The Times*, 2006

20   Goals and Objectives, Maggie's Centre, maggiescentres.net, 2006

21   Lesley Fallowfield et al., 'Addressing the psychological needs of the conservatively treated breast cancer patient: discussion paper', *Journal of the Royal Society of Medicine*, 1987

22   Dennis Barker, 'Dame Cicely Saunders – obituary', the *Guardian*, 2005

23   Robert Houston and Gary Null quoted by cancertutor.com, 2007

24   Impossible Dream – Why The Cancer Industry Is Committed To *Not* Finding a Cure, and If They Do Run Across One, Suppressing It! Altcancer.com, 2006

25   Mark Lipsman, A holistic approach to cancer, goodhealthinfo. net, 2006

26   Thomas J Wheeler PhD, Cancer and AIDS therapies. A Scientific Look at Alternative Medicine, University of Louisville School of Medicine, 2006

27   Aspirin 'prevents cancer', BBC News report, 2003

28   Peter J Bryant, Natural Products, Biology and Conservation hypertext book, School of Biological Sciences, University of California, 2002

29   Robin Pagnamenta, Merck takes dose of alternative medicine, *The Times*, 2006

30   Sarah-Jane Templeton, 'Cancer treatment – privilege of the rich?' *The Times*, 2005

31   essiaceu.com, 2007

32   cancerfightingstrategies.blogspot.com. 2006

33   Alternative and Self-Help Cancer Therapy Programs For People and Animals, shirleys-wellness-café.com, 2006

34   M D Holmes et al., 'Exercise after treatment may improve survival and reduce recurrence', *Journal of the American Medical Association*, 2005

35   E Ernst and K Schmidt, 'Assessing websites on complementary and alternative medicine for cancer,' *Annals of Oncology*, 2004

36 Agnes Tiller, Alternative cancer treatment & holistic healing approaches: cures and remission testimonies, healingcancernaturally.com, 2007

37 cancerfightingstrategies.blogspot.com, 2006

38 *Icon* Magazine, Issue 3, Volume 5, CANCERactive, canceractive.com, 2006

39 Hoxsey Clinic, Alternative Cancer Clinic – Biomedical Clinic, Tijuana, Mexico, cancure.org, 2007

40 Lesley Garner, 'Formula that gave new life', *The Daily Telegraph*, 2002

41 Marian Segal, 'Court Says to Cancel the Cancell', FDA Consumer magazine, 1993

42 Hulda Regehr Clark, *The Cure for All Cancers*, New Century Press, 1993

43 Barrie R Cassileth and Gary Deng, 'Complementary and Alternative Therapies for Cancer', *The Oncologist*, 2004

44 Jane Sen and Rosy Daniel, *Eat to Beat Cancer – A Nutritional Guide with 40 delicious recipes*, HarperCollins, 2003

45 Professor Jonathan Waxman, 'Shark Cartilage in the Water', *BMJ*, 2006

46 William T Jarvis, 'Helping your Patients deal with Questionable Cancer Treatments', *CA: A Cancer Journal for Physicians*, 1986

47 Healing your Body with the Gerson Therapy, The Gerson Institute, 2006

48 Incidence of the major cancers 2004, Office of National Statistics, statistics.gov.uk, 2006

49 Kevin Trudeau, *Natural Cures 'they' don't want you to know about*, Perseus Books, 2005

50 Order form for the 'Norwalk Juicer', Norwalk Juicer California, nwjcal.com, 2006

51 2006 The Gerson Diet, Complementary and Alternative Medicines in the Cancer Field, cam-cancer.org, 2006

52 The Prince of Wales, speech at conference: Complementary Therapies and Cancer Care – A Research Symposium, princeofwales.gov.uk, 2004

53 Peter Moran, Dr Max Gerson and his fifty cases (How sincere practitioners can get it wrong), Alternative medicine and cancer, cancerwatcher.com, 2006

54 Gerson Therapy, BC Cancer Agency, 2006

55 Stephen Barrett MD and Victor Herbert MD, Questionable Cancer Therapies, Quackwatch.org, 2001

56 J W Eisele and D T Reay, 'Deaths related to coffee enemas', *JAMA*, 1980

57 J P Davignon et al., 'Pharmaceutical assessment of amygdalin (Laetrile) products', *Cancer Treatment Reports*, 1978

58 S Milazzo et al., *Laetrile Treatment for Cancer*, Cochrane Review, John Wiley and Sons, 2006

59 C G Moertel et al., 'Clinical trial of amygdalin in the treatment of human cancer', *New England Journal of Medicine*, 1982

60 Laetrile (amygdalin, Vitamin B17): side effects, cancerhelp. org.uk

61 Barron H Lerner, 'McQueen's legacy of Laetrile', *New York Times*, 2005

62 William Donald Kelley, Cancer Cure Suppressed, whale.to/cancer/kelley

63 Ralph W Moss, William Donald Kelley, Townsend Letter for Doctors and Patients, 2005

64 Barron H Lerner, *When Illness Goes Public*, The Johns Hopkins University Press, 2006

65 Rene Caisse, *I was Canada's Cancer Nurse*, booklet undated but widely available on the internet, e.g at essiacinfo.org

66 Barrie R Cassileth, *Alternative medicine handbook: a complete reference guide to alternative and complementary therapies*, WW Norton & Co, 1998

67 C Dold, 'Shark Therapy', *Discover*, 1996

68 Shark Cartilage powder product information, hwize.com, 2006

69 Gary K Ostrander et al., Shark Cartilage, cancer and the growing threat of pseudoscience, Cancer Research, American Association for Cancer Research, 2004.

70 Shark Cartilage, Cancer Research UK, cancerhelp.org, 2006

71 Maggie Fox, 'Shark Cartilage no help against lung cancer': study, *Reuters*, 2007

72 'Shark Cartilage Cancer "cure" shows danger of pseudoscience', Johns Hopkins University press release, *Science Daily*, 2004

73 'Operation Cure.all' Nets Shark Cartilage Promoters: Two Companies Charged with Making False and Unsubstantiated Claims for Their Shark Cartilage and Skin Cream as Cancer Treatments, Press Release, US Federal Trade Commission, 2000

74 Jean-Michel Cousteau, 'Sharks at Risk', *Ocean Adventures* TV series, PBS, 2006

75 Patricia Sullivan, 'Biologist Ransom A Myers, 54; warned of overfishing in oceans', *Washington Post*, 2007

76 Stephen Barrett MD, Book Review, American Cancer Society's Guide to Complementary and Alternative Cancer Methods, Quackwatch.org 2001

77 Gabriella M D'Andrea, 'Use of antioxidants during chemotherapy and radiotherapy should be avoided', *CA: A Cancer Journal for Physicians*, 2005

78 Jonathan Waxman, 'Shark Cartilage in the Water', *British Medical Journal*, 2006

79 Probable Cause Affidavit, State of Indiana County Court, 1993

80 blacksalveinfo.com, 2006

81 curezone cancer forum, curezone.com, 2006

82 Penny Brohn, *Gentle Giants: the powerful story of one woman's fight against breast cancer*, Century, 1987

83 Dr Robert Buckman and Karl Sabbagh, *Magic or Medicine: an investigation into healing*, Pan Books, 1993

84 Alternative cancer treatments – comparison and testing, alternativecancer.us, 2006

85 Gloria Hunniford, *Next to You – Caron's Courage Remembered by Her Mother*, Penguin, 2006

86 Brandon Bays, thejourney.com

87 Brandon Bays, *The Journey*, HarperCollins, 1999

88 'Think Twice about Taking The Journey', *Newsletter of the British False Memory Society*, 2005

89 *Instant Calm with Caron Keating*, Universal Pictures Video, 2002

90 Gloria Hunniford, *Next to You – Caron's Courage Remembered by Her Mother*, Penguin Books, 2005

91 Sue Mott, 'Bazza's biggest race', *Melbourne Age*, 2002

92 Survival rates for stomach cancer, cancerhelp.org, 2006

93 Tony Sheldon, 'Dutch doctors suspended for use of complementary medicine', *British Medical Journal*, 2006

94 Millecam doctors suspended, Bad Science, 2007, badscience.net

95 Canadian Cancer Research Group, ccrg.com, 2007

96 Nate Hendley, 'Cures of Last Resort', *Eye Weekly*, 2000

97 Alan Fryer, *Dr Hope*, W-Five, Canadian Television, 2006

98 CCRG pursues CTV W-Five, Canadian Cancer Research Group, ccrg.com, 2007

99 *Bogus Cancer Cures, Week In Week Out*, BBC Wales, 2003

100 Louise Mclean, 'Persecution of alternative practitioner treating cancer in the UK', *Institute of Complementary Medicine Journal*, 2005

101 Press Release: Former vicar punished for illegally selling unlicensed medicines, Medical & Healthcare Products Regulatory Agency, 2006

102 *The Anatomy of Hope*, Pocket Books, 2004

103 nueracarecentre.com, 2006

104 Sue Ellen Hinde, 'Cruel end to dash for cancer "cure"', *Weekend Australian*, 2006

105 'Court finds cancer sufferers exploited under the RANA system', press release, Australian Competition and Consumer Commission, accc.gov.au, 2007

106 Edmund Tadros and agencies, 'Australian cancer patient killed by quack doctor's poison, Thai police say', *Sydney Morning Herald*, 2006

107 Martin Daly and Leonie Wood, 'Dr Ozone's long history of preying on the terminally ill', *Melbourne Age*, 2006

108 Ryke Geerd Hamer, Introduction to Dr Hamer's New Medicine, newmedicine.ca

109 Summary of Trial Proceedings, Hamer v France, sim.law.uu.nl, 2006

110 We mourn for Michaela, ariplex.com/ama/amamiche.htm, 2007

111 'William Jarvis, Helping Patients Deal with Questionable Cancer Treatments', *Ca: A Cancer Journal for Clinicians*, 1986

Chapter 8: Fads, Fashions and Faking it

1 TV interview with Dr Jacob Teitelbaum, fatigue specialist, *Good Morning Texas*, WFAA, 2006

2 Parasites: a myriad of parasitical creatures call the body home! Parasitecleanse.com, 2007

3 Sir Thomas Browne, *Letter to A Friend*, pamphlet, 1690, can be read in full at website penelope.uchicago.edu/letter/letter.html

4 Morgellons Research Foundation, Morgellons.org, 2007

5 Michael Mason, 'Is it disease or delusion?' *New York Times*, 2006

6 Elaine Monaghan, 'All in the Head?' *Times*, 2006

7 Pimozide, patient.co.uk, 2007

8 Elaine Monaghan, 'All in the Head?' *Times*, 2006

9 Erin Allday, 'CDC to investigate mystery disease', *San Francisco Chronicle*, 2006

10 skinparasites.com, 2007

11 Morgellons Treatment, safe2use.com

12 American Thyroid Association's statement on Wilson's Syndrome, American Thyroid Association, thyroid.org, updated 2005

13 wilsonssyndrome.com, 2007

14 Dr Peatfield's Clinics 2007, thyroid-disease.org.uk, 2007

15 Dr Barry Durrant-Peatfield, *The Great Thyroid Scandal and How to Survive It*, Hammersmith Press, 2006

16 'GMC suspends doctor over alternative thyroid treatment', *Independent*, 2001

17 Interim Orders, Gordon Skinner, General Medical Council, 2007

18 General Medical Council Abolition Petition gets Government Approval, medicalnewstoday.com, 2007

19 Arthur J Barsky, *Worried Sick: Our Troubled Quest for Wellness*, Little Brown, 1988

20 Scott Haig MD, 'Doctors Without Dollars', *Time Magazine*, 2007

21 Arthur J Barsky and Emily C Deans, *Stop being your symptoms and start being yourself*, HarperCollins, 2006

22 Carrie McClaren, 'The history of psychosomatic illness: an interview with Edward Shorter', *Stay Free!* magazine, 2003

23 Barbara Ehrenreich and Deirdre English, *Complaints and Disorders: The sexual politics of sickness*, Glass Mountain Pamphlet no 2, The Feminist Press, 1974

24 Carrie McClaren, 'The history of psychosomatic illness: an interview with Edward Shorter', *Stay Free!* magazine, 2003

25 Edward Shorter, *From Paralysis to Fatigue – a History of Psychosomatic Illness in the Modern Era*, Free Press, 1992

26 L A Page and S Wessely, 'Medically unexplained symptoms: exacerbating factors in the doctor-patient encounter', *Journal of the Royal Society of Medicine*, 2003

27 John Rigg, 'Disabling attitudes? Public perspectives on disabled people', in *British Social Attitudes: the 23rd report – Perspectives in a Changing Society*, National Centre For Social Research, Sage, 2007

28 Elaine Showalter, *Hystories – Hysterical Epidemics and Modern Culture*, Picador, 1997

29 Post Nasal Drip, sinuswars.com 2007

30 What is ME? Action for ME, 2007

31 About the 25% ME group, 25megroup.org, 2007

32 Duncan Chambers et al., 'Interventions for the treatment, management and rehabilitation of patients with chronic fatigue syndrome/myalgic encephalomyelitis: an updated systemic review', *Journal of the Royal Society of Medicine*, 2006

33 Rachel Anderson discussing her book *This Strange New Life*, *Woman's Hour*, BBC Radio 4, 2006

34 Esther Rantzen, 'Saved from a Living Death', *Daily Mail*

35 Neil C Abbot and Vance Spence, Chronic fatigue syndrome: a response to the Lancet review by Prins et al., ME Research UK, 2007

36 Judith B Prins et al., 'Review: Chronic Fatigue Syndrome', *The Lancet*, 2006

37 Anthony J Pinching, 'AIDS and CFS/ME: a tale of two syndromes', *Clinical Medicine*, 2003

38 Michael Sharpe and David Wilks, 'Fatigue, ABC of psychological medicine', *British Medical Journal*, 2002

39 Judith B Prins et al., 'Review: Chronic Fatigue Syndrome,' *The Lancet*, 2006

40 NICE guidelines amended, press release, Action for ME, 2007

41 Professor Simon Wessely, 'Chronic Fatigue Syndrome – trials and tribulations', *Journal of the American Medical Association*, 2001

42 Professor Simon Wessely, 'Something old, something new, something borrowed, something blue', lecture delivered at Gresham College London, 2006

43 Peter White et al., 'Chronic Fatigue Syndrome or Myalgic Encephalomyelitis', *British Medical Journal*, 2007

44 Richard E Harris et al., 'How do we know that the pain of Fibromyalgia is "real"?' *Current Pain and Headache Reports*, 2006

45 What is Fibromyalgia? US Department of Health and Human Services Public Health Service, 2007

46 Professor Nortin M Hadler, Fibromyalgia: could it be in your mind? Cutting Edge Reports, rheuma21st.com, 2001

47 Jerome Groopman MD, 'Hurting all over: fibromyalgia', *New Yorker*, 2000

48 Lewis Krauskopf, Drug Nearing Approval for Mysterious Health Condition, Reuters Health Information, 2007

49 Reported Cases of Lyme Disease by Year, United States 1991-2005, US Centers for Disease Control, 2006

50 Frequently Asked Questions on Lyme Borreliosis, UK Health Protection Agency, 2006

51 Epidemiology of Lyme borreliosis, UK Health Protection Agency, 2006

52 Dr Allen Steere, Lyme Disease – questions and answers, Massachusetts General Hospital/Harvard Medical School, 2003

53 BADA-UK Statement, Ho-Yen Lyme Disease Article, One Click Group, theoneclickgroup.co.uk, 2006

54 Dr Darrel Ho-Yen, 'Lyme Disease – Let's Dispel the Myths', *ME Essential*, 2006

55 Press release, LIA Foundation Hosts the First Conference Dedicated to Educating on the Link Between Lyme disease and Autism, Lyme Induced Autism Foundation, liafoundation.org, 2007

56 J D Cooper and H M Feder Jr, 'Inaccurate information about Lyme disease on the Internet', *Journal of Paediatric Infectious Diseases*, 2004

57 Professor Brian L Duerden, Unorthodox and unvalidated laboratory tests in the diagnosis of Lyme Borreliosis and in relation to medically unexplained symptoms, Department of Health, 2006

58 Controversies in Lyme Borreliosis, Health Protection Agency, 2006

59 Gary P Wormser et al., 'The Clinical Assessment, Treatment, and Prevention of Lyme Disease, Human Granulocytic Anaplasmosis, and Babesiosis: Clinical Practice Guidelines by the Infectious Diseases Society of America', *Clinical Infectious Diseases*, 2006

60 Diagnosis of Lyme Borreliosis, Health Protection Agency, 2007

61 Dr Allen C Steere et al., 'The overdiagnosis of Lyme disease', *Journal of the American Medical Association*, 1993

62 David Grann, 'Stalking Dr Steere over Lyme Disease', *New York Times Magazine*, 2001

63 Edward Shorter, *From Paralysis to Fatigue – a History of Psychosomatic Illness in the Modern Era*, Free Press, 1992

64 *Sold on Syndromes*, CBS Radio, 1998

65 Gina Kolata, 'Lyme Disease is Hard to Catch and Easy to Halt, Study Finds', *The New York Times*, 2001

66 Sandra Cabot MD, Gall bladder disease and the gall bladder flush, liver-doctor.com, 2007

67 Hulda Clark, *The Cure for All Diseases*, B. Jain publishers, 2005

68 Christiaan W Sies and Jim Brooker, 'Could these be gallstones?' *The Lancet*, 2005

69 Andreas Moritz, Books by Andreas, Ener-Chi Wellness Centre, ener-chi.com, 2007

70 Testimonials, colon cleansing, drnatura.com, 2007

71 Richard Anderson, Should I Colon Cleanse? cleanse.net, 2006

72 Edward Uthman MD, Mucoid Plaque, Quackwatch, 1998

73 Richard Anderson, My Most Frequently Asked Questions about Colon and Internal Cleansing, cleanse.net, 2007

74 Black Psyllium, Natural Medicines Comprehensive Database, 2006

75 Minerals and Ores, Sorptive Minerals Institute, sorptive.org 2007

76 drnatura.com, 2007

77 Professor Edzard Ernst, 'Colonic irrigation and the theory of autointoxication: a triumph of ignorance over science', *Journal of Clinical Gastroenterology*, 1997

78 Dr Ben Goldacre, 'A Menace to Science', the *Guardian*, 2007

79 Detox press release, Sense about Science, 2006

80 Dr John Briffa, 'What's in your basket?' *Observer Food and Drink* magazine, 2007

81 The Detox body wrap, Shape Changers, Shapechangers.co.uk

82 Crystal Spring Detoxifying Footpads, crystalspring.co.uk, 2007

83 Suellen Hinde, 'Diet fad threat to kids', *Herald Sun*, 2007
84 Anita Bean, *Carol Vorderman's Detox for Life*, Virgin Books, 2002
85 Adverse Reactions, Vitamin A, Natural Medicines Comprehensive Database, 2006
86 Detox and re-energise your life, boots.com, 2007
87 Products: Clarins Expertise 3P, uk.clarins.com, 2007
88 Detox press release, Sense about Science, 2006
89 *The Truth About Food: Detox*, BBC2, 2007
90 Ray Moynihan and David Henry, 'The fight against disease mongering: generating knowledge for action', *PLoS Medicine*, Public Library of Science, 2006

Chapter 9: It Works for Me

1 M Lamar Keene and Allen Spragg, *The Psychic Mafia*, Prometheus Books, 1998
2 Ann Callaghan, Indigo Essences, indigoessences.com, 2007
3 S E Harris, 'Behind the Label: Dr Gillian McKeith's living food love bar', *The Times*, 2004
4 Aromatherapy Pamper Day 2007, Sir John Deane's College, sjd.ac.uk, 2007
5 Julie Burchill, 'Buswoman's holiday', the *Guardian*, 2000
6 Immune System Tonic, Aromatherapy Bath Oil, edirectory.co.uk, 2007
7 Tracy-Ann Oberman, 'I want my body back', the *Guardian*, 2007
8 Helen Lederer, 'Cultureshock', the *Guardian*, 2007
9 Sheila Pulham, 'Massage and a bottle', the *Guardian*, 2007
10 LaStone with Sarah Kirby Therapies, Brighton and Hove, Sussex UK, sarahkirbytherapies.com, 2007
11 Therapy A-Z, Wood Green Complementary Health Centre, woodgreen-health.co.uk, 2007
12 Nick Ferrin, The homeopathic treatment of autism, The Autism File, autismfile.com, 1999
13 Professor Raymond Tallis, 'Longer, Healthier, Happier? Human needs, human values and science', Sense about Science lecture, 2007
14 Darian Leader and David Corfield, *Why Do People Get Ill?* Hamish Hamilton, 2007
15 Dr Margaret McCartney, 'Reclaiming the placebo', *Prospect Magazine*, 2007
16 Ben Goldacre, 'A tonic for sceptics – Bad Science', the *Guardian*, 2005
17 Catherine Zollman and Andrew Vickers, *ABC of Complementary Medicine*, BMJ Books 2000
18 Helios Double Helix Kits, Helios Remedies, helios.co.uk
19 You and the Bach Original Flower Remedies, Bach Flower Remedies Ltd, 2005
20 Michael Odell, 'Jerry Hall', *Observer*, 2007
21 Benedict Carey, 'Hard to Swallow', *Los Angeles Times*, 2001
22 Heida Knapp Rinella, Eight, 8 ounce glasses, consuming proper amount of water keeps mind, body sharp, Chet Day's Health and Beyond, chetday.com, 2007
23 Dr Fereydoon Batmanghelidj, *Your Body's Many Cries for Water – a Revolutionary Natural Way to Prevent Illness and Disease and Restore Good Health*, Tagman Press, 2004
24 Eight Glasses, Urban Legends Reference Pages, snopes.com, 2007
25 The UK Market, Highland Spring, highlandspring.com, 2007
26 Rebecca Smithers, 'Fizzy drinks giving way to water and juice', the *Guardian*, 2007

27 Press release: Pennine Spring Water, Britvic Soft Drinks Ltd, 2006

28 Recommended Dietary Allowances, Food and Nutrition Board, National Academy of Sciences, National Research Council, Reprint and Circular Series No 122, 1945

29 'Does drinking two litres of water a day improve your skin?' *The Truth About Food*, BBC TV, 2007

30 *Need for 8 glasses of water a myth: nutrition experts*, CBC News, cbc.ca, 2003

31 Ann Grandjean et al., 'The effect of caffeinated, non-caffeinated, calorific and non-calorific beverages on hydration', *Journal of the American College of Nutrition*, 2000

32 'Tea "healthier" drink than water', BBC News, 2007

33 Dr Heinz Valtin, 'Drink at least eight glasses of water a day. Really? Is there scientific evidence for "8 x 8"?' *American Journal of Regulatory, Integrative and Comparative Physiology*, American Physiological Society, 2002

34 Rukmini Callimachi, 'Bitter Medicine', the *Guardian*, 2007

35 Jeff Koinange, *In Gambia, AIDS cure or false hope?* CNN, 2007

36 *Gambia's UN Envoy 'is expelled'*, BBC News, 2007

37 Barry L Beyerstein, Why bogus therapies often seem to work, Quackwatch, 2003

38 Karl Sabbagh, 'The Psychopathology of Fringe Medicine', *Skeptical Inquirer*, CSICOP, 1985

39 Applied Kinesiology in Oxford UK, Helix House Natural Health Centre, helixhouse.co.uk, 2007

40 Diana Sheppard, 'Don't be a Grumble Guts!' practice advertisement in *The Spark*, 2007

41 James Randi, Applied Kinesiology, An Encyclopedia of Claims, Frauds and Hoaxes of the Occult and Supernatural, Randi.org/encyclopedia, 2007

42 Ray Hyman, 'The mischief making of ideomotor action', *The Scientific Review of Alternative Medicine*, 1999

43 Herman H Spitz and Yves Marcuard, quoting Chevreul's reports on the Mysterious Oscillations of the hand-held pendulum (Lettre M. Ampere, *Sur une classe particuliere de mouvements musculaires*, 1833) *Skeptical Inquirer*, CSICOP, 2001

44 Chevreul's Pendulum, everything2.com, 2007

45 H-J Staele et al., 'Double-Blind Study on Materials Testing with Applied Kinesiology', *Journal of Dental Research*, 2005

46 R Pothmann et al., 'Evaluation of Applied Kinesiology in Nutritional Intolerance of Childhood', *Research in Complementary Medicine*, 2001

47 Benedict Carey, 'Do you Believe in Magic?' *New York Times*, 2007

48 Phillips Stevens Jr, 'Magical Thinking in Complementary and Alternative Medicine', *Skeptical Inquirer*, 2001

49 John and Henry Harries, National Trust Visitor Centre display, Pumpsaint, Carmarthenshire, 2007

50 Aura Soma®, 2007 Price list, footprintsandpathways.co.uk

51 Sir James George Fraser, *The Golden Bough: A Study in Magic and Religion*, (abridged) Penguin Books, 1996 edition

52 *Homeopathy, Complementary therapies – check it out*, Boots information booklet, 2006

53 Karl Sabbagh, 'The Psychopathology of Fringe Medicine', *Skeptical Inquirer*, CSICOP, 1985

54 Peter Skrabanek and James McCormick, *Follies and Fallacies in Medicine*, Buffalo, NY, Prometheus, 1990

55 Tamar Nordenberg, 'The Healing Power of Placebos', *FDA Consumer* magazine, 2000

56 Jon-Kar Zubieta, 'Placebo Effects Mediated by Endogenous Opioid Activity on μ-Opioid Receptors', *Journal of Neuroscience*, 2005

57 Aaron Vallance, 'Something out of Nothing – the placebo effect', *Advances in Psychiatric Treatment*, 2006

58 Henry K Beecher, 'The Powerful Placebo', *Journal of the American Medical Association*, 1955

59 Asbjorn Hrobjartsson and Peter C Gotzshe, 'Is the Placebo Powerless? – Update of a systemic review with 52 new randomised trials comparing placebo with no treatment', *Journal of Internal Medicine*, 2004

60 Dylan Evans, *Placebo – mind over matter in modern medicine*, HarperCollins, 2005

61 J Bruce Moseley et al., 'A controlled trial of arthroscopic surgery for osteoarthritis of the knee', *New England Journal of Medicine*, 2002

62 Andrew Moore and Henry McQuay, *Bandolier's Little Book of Making Sense of the Medical Evidence*, Oxford University Press, 2006

63 Brian Reid, 'The Nocebo Effect – Placebo's Evil Twin', *Washington Post*, 2002

64 Arthur J Barsky et al., 'Nonspecific medical effects and the nocebo phenomenon', *Journal of the American Medical Association*, 2002

65 Ted J Kaptchuk et al., 'Sham device versus inert pill: randomised controlled trial of two placebo treatments', *British Medical Journal*, 2006

66 Anton J M de Craen, 'Effect of Colour of Drugs: a systematic review of perceived effects of drugs and of their effectiveness', *British Medical Journal*, 1996

67 Edzard Ernst, Mind over Matter? Arthritis Research Campaign, arc.org.uk, 2006

68 David Peters ed., *Understanding the Placebo Effect in Complementary Medicine*, Churchill Livingstone, 2001

69 Philip Welsby, 'Passionate Medicine – or magical thinking?' *healthmatters*, 2006

70 Alun Anderson, 'Physician, fool thyself', *Fortune Magazine*, 2006

71 David Spiegel, 'Placebos in Practice', *British Medical Journal*, 2004

72 David Wootton, *Bad Medicine: Doctors Doing Harm since Hippocrates*, Oxford University Press, 2006

73 Margaret McCartney, 'Reclaiming the placebo', *Prospect Magazine*, 2007

74 Zelda Di Blasi and David Reilly, 'Placebos in medicine: medical paradoxes need disentangling', *British Medical Journal*, 2005

75 Dylan Evans, *Placebo – mind over matter in modern medicine*, HarperCollins, 2005

Chapter 10: The Healthy Sceptic

1 Prof Raymond Tallis, 'Longer, Healthier, Happier? Human needs, human values and science', Sense about Science lecture 2007

2 Dan Hurley, *Natural Causes – death, lies, and politics in America's vitamin and herbal supplement industry*, Broadway Books, 2006

3 E Ernst and A R White, 'The BBC Survey of Complementary Medicine Use in the UK', *Complementary Therapies in Medicine*, 2000

4 P Harris et al., 'Use and expenditure on complementary medicine in England, a population-based survey', *Complementary Therapies in Medicine*, 2001

5 'Natural medicine on trial: The trouble with herbs', Jeremy Laurance, *Independent*, 2007

6 Christopher Smallwood et al., 'The Role of Complementary Medicine in the NHS', Fresh Minds for The Prince's Foundation for Integrated Health, 2005

7 Mark Henderson and Fran Yeoman, 'NHS must audit spending on alternative therapy, MPs say', *The Times*, 2006

8 Caroline Flint, letter to Member of Parliament reproduced on Prof David Colquhoun's Quack Page, 2006

9 Julia Stuart, 'Children receive spiritual healing on the NHS', *Independent on Sunday*, 2005

10 'Spiritual Healers' using up scarce NHS resources, National Secular Society, secularism.org.uk, 2007

11 University College Hospital NHS Healer, Cancer/Haematology unit, advertisement, National Federation of Spiritual Healers, nfsh.org.uk, 2007

12 Lee Glendinning, 'Hospital kitchens fail to meet basic hygiene levels', the *Guardian*, 2007

13 *Malnutrition 'time bomb' warning*, BBC News, 2005

14 Peter Hain MP, *Speech to Foundation for Integrated Health*, 2005

15 Sarah Ross et al., 'Homeopathic and herbal prescribing in Scotland', *British Journal of Pharmacology*, 2006

16 Acupuncture helps more patients, media release, NHS Highland, 2007

17 The Prince's Foundation for Integrated Health, Integrated Health: the best of healthcare for the whole person, 2006

18 *Charles, The meddling prince*, *Dispatches*, Channel 4, 2007

19 Sir Michael Peat, Letter to Dispatches, princeofwales.gov.uk, 2007

20 Gerard Weissman, 'Homeopathy: Holmes, Hogwarts, and the Prince of Wales', Editorial, *FASEB Journal*, Federation of American Societies for Experimental Biology, 2006

21 Professor Michael Baum, contributor, *Does Alternative Medicine Work?* Today programme, BBC Radio 4, 2006

22 Prof Michael Baum and twelve others, Use of 'alternative ' medicine in the NHS, open letter to NHS trusts, 2006

23 Professor Gustav Born and six others, letter to NHS trusts calling for homeopathy boycott, 2007

24 Graham Clews, 'PCTs consider alternatives to homeopathic hospitals', *Health Service Journal*, 2007

25 Dr Peter Fisher, The Royal London Homoeopathic Hospital needs your support, letter to supporters, 2007

26 Dr Peter Fisher, Open Letter, Closed Minds, Comment is Free, Guardian Unlimited, guardian.co.uk, 2007

27 Kate Maxwell, 'Homeopathy is worse than witchcraft – and the NHS must stop paying for it', *Daily Mail*, 2007

28 West Kent PCT press release, 'West Kent PCT ends funding for homeopathy', 2007

29 *The NHS Directory of Complementary and Alternative Practitioners*, NHS Trusts Association, 2006

30 Complementary Alternatives, NHSTA, complementaryalternatives.com, 2007

31 Mark Henderson, 'Faith-based degree "damages science"', *The Times*, 2007

32 David Colquhoun, 'Science degrees without the science', *Nature*, 2007

33 Provisional figures for 2006 entry, Universities and Colleges Admissions Service, ucas.ac.uk

34 University of Newcastle medicine degree course requirements, Universities and Colleges Admissions Service, ucas.ac.uk

35 Course finder, BSc (Hons) Complementary Medicine and Health Sciences, Faculty of Health & Social Care, University of Salford, 2007

36 Brian Isbell, speech to prospective students, University of Westminster CAM course open day, 2006

37 Cost based on 2006 student numbers and HEFCE funding bands

38 Jonathan Swift, *Gulliver's Travels*, 1726

39 Daniel Elkan, 'Clinical excellence', the *Guardian*, 2007

40 Prof Rory Coker, Distinguishing Science from Pseudoscience, Quackwatch, Quackwatch.org 2001

41 Richard P Feynman and Ralph Leighton, *Surely You're Joking Mr Feynman! Adventures of a Curious Character*, Vintage, 1992

42 Raymond Tallis, 'Longer, Healthier, Happier? Human needs, human values and science', Sense about Science Lecture, 2007

43 Prof Richard Dawkins, Foreword to John Diamond's *Snake Oil and Other Preoccupations*, Vintage, 2001

44 Phil Fontanarosa and George Lundberg, 'Alternative Medicine meets Science', *JAMA*, 1998

45 Thomas J Wheeler, 'A Scientific Look at Alternative Medicine', Department of Biochemistry and Molecular Biology, University of Louisville School of Medicine, 2005

46 National Service Framework for Mental Health: Modern Standards and Service Models, UK Department of Health, 1999

47 John Diamond, *Snake Oil and Other Preoccupations*, Vintage, 2001

48 Tricia Greenhalgh, *How to Read a Paper*, BMJ Books, 2000

49 Andrew Moore and Henry McQuay, *Bandolier's Little Book of Making Sense of the Medical Evidence*, Oxford University Press, 2006

50 Edzard Ernst, 'Medicine Man', the *Guardian*, 2004

51 Kathy Sutton and Naomi Stevenson, *Public health, private wealth*, Fellows' Associates, 2005

52 Kathy Sutton, 'Missing the Point', *healthmatters* magazine, healthmatters publications, 2005

53 Keith I Block and Wayne B Jonas, 'Top of the Hierarchy' Evidence for Integrative Medicine: What are the best strategies? Integrative Cancer Therapies, 2006

54 Research in Homeopathy – comments on homeopathy research, British Homeopathic Association, 2007

55 Wallace Sampson and Lewis Vaughn eds., *Science meets alternative medicine: what the evidence says about unconventional treatments*, Prometheus Books, 2000

56 Thomas J Wheeler, 'A Scientific Look at Alternative Medicine', Department of Biochemistry and Molecular Biology, University of Louisville School of Medicine, 2005

57 Kimball C Atwood, 'Prior probability: the dirty little secret of "evidence based" alternative medicine', paper delivered at the 11th European Skeptics Congress, London, 2003

58 Professor Linda Franck et al., 'Should NICE evaluate complementary and alternative medicine?' *British Medical Journal*, 2007

59 Professor David Colquhoun, 'Should NICE evaluate complementary and alternative medicine?' *British Medical Journal*, 2007

60 Angela Coulter, 'Killing the goose that laid the golden egg?' *British Medical Journal*, 2003

61 J P McLaughlin, 'On looking and leaping and critical thinking: Boulder critical thinking workshop', *Skeptical Inquirer*, 1995

Index